HOLIDAY CRUISING IN THE NETHERLANDS

The Holiday Cruising series

HOLIDAY CRUISING ON INLAND WATERWAYS
by Charles Hadfield and Michael Streat

HOLIDAY CRUISING ON THE THAMES
by E. and P. W. Ball

HOLIDAY CRUISING ON THE BROADS AND FENS
by Lewis Edwards

HOLIDAY CRUISING IN IRELAND
by P. J. G. Ransom

HOLIDAY CRUISING IN FRANCE
by Gerard Morgan-Grenville

HOLIDAY CRUISING IN THE NETHERLANDS
by John Oliver

The author's yacht *Puffdragon* sailing in a Dutch canal

HOLIDAY CRUISING IN THE NETHERLANDS

with plates, maps and diagrams

JOHN OLIVER

DAVID & CHARLES
NEWTON ABBOT LONDON

The author has done his best to make sure that the information given in this book was correct when it went to press, but neither he nor the publishers can hold themselves responsible for the consequences of any mistakes.

0 7153 6488 X

Set in 11 pt Times New Roman 2 pt leaded
and printed in Great Britain
by Latimer Trend & Company Ltd Plymouth
for David & Charles (Holdings) Limited
South Devon House Newton Abbot Devon

Contents

	List of Illustrations	*page* 7
	Introduction	9
CHAPTER		
1	WHY HOLLAND?	11
2	PREPARATIONS	14
	Sailing Across	14
	Trailed Boats	15
	Hiring a Boat in Holland	19
	Formalities	22
3	THE CRUISING GROUNDS	23
	The Schelde and the Veersemeer	24
	The Ijsselmeer	39
	Friesland	49
	Friesian Islands	62
	Western Holland	66
	The Great Rivers	81
	Through Routes	94
	Waterways Not Recommended	97
4	LIVING ON THE CANALS	100
	Climate	100

	Shops	*page* 100
	Eating Out	101
	Evening Entertainment	101
	Other Canal Users	102
	Repairs	105
	Houseboats	105
	Fuel	106
	Cycle Hire	106
	Public Transport	109
5	NAVIGATION	110
	Bridges and Locks	110
	Buoys	113
	Beacons	115
6	CANAL REGULATIONS	123
	Vaarreglement	123
	Algemeen Reglement van Politie	124
	Reglement van Politie voor de Rijnvaart	125
7	OTHER INFORMATION	129
	Charts and Books	129
	Tourist Information	133
	British Consulates	134
	Annual Events	134
	Ferries	135
	Glossary of Useful Words	136
	Appendix A Weights and Measures	141
	Appendix B Mast Lowering Gear	144
	Acknowledgements	147
	Index	148

List of Illustrations

PLATES

The author's yacht *Puffdragon* sailing in a Dutch canal — *frontispiece*
A typical windmill at sunset (*Clive Hodsdon*) — *page* 17
Late-Gothic town hall at Gouda — 17
The magnificent Zeelandbrug — 18
Storm gates at Krimpen a/d Ijssel — 18
The author tries his hand at the 'national pastime' (*Nick Brown*) — 35
One of Holland's rare deep locks — 35
Waterbus in Amsterdam (*Nick Brown*) — 36
River police boat at Muiden — 36
The jachthaven at Vlissingen — 53
A quiet mooring on the Veersemeer — 53
Holiday cruising, traditional Dutch style — 54
A *zeeschow* reaching across the Ijsselmeer — 54
The lighthouse at Marken — 71
New locks in the dyke surrounding Markerward — 71
The harbour at Volendam — 72
The entrance to the old harbour at Enkhuizen — 72
A typical Friesian farmhouse (*Nick Brown*) — 89
A modern cantilever bridge — 89
Yachts tucked in amongst the trawlers at Oudeschild — 90
The author's yacht baulked by a bridge in Groningen — 90

List of Illustrations

The old and the new. Scene on the River Ijssel *page* 107
Motor boats in the jachthaven at Zutphen 107
Entering Amsterdam from the south (*Clive Hodsdon*) 108
Ferry in Amsterdam (*Nick Brown*) 108
Puffdragon tied up by the roadside in Makkum 117
Houseboat on the River Vecht 117
A dredger at work 118
Quite a small barge by Dutch standards (*Clive Hodsdon*) 118
A typical cantilever bridge 119
Passing through a swing bridge 119
A *tjackshalt* 120
Radboud Castle (*Nick Brown*) 120

MAPS

Waterways navigable by yachts with fixed masts (southern part) 20
Waterways navigable by yachts with fixed masts (northern part) 21
Waterways of southern Holland 25
The Ijsselmeer 41
The canals and lakes of Friesland 51
The Friesian islands 64
Waterways of western Holland 68
The great rivers 83

DRAWINGS

Buoys to be found on the inland waterways 114
Beacons used in the Netherlands 116
Speed conversion graph 143
Mast lowering gear 145

All photographs are by the author unless otherwise acknowledged.

Introduction

Less than twenty-four hours sailing from my own river, the Medway, and I can enter a new world. A world where sailing is not thought eccentric or the province of the rich. A world where the grass is greener, the air is cleaner and pollution is a hazard yet to be encountered. A world where small cruising dinghies, weekend runabouts, family cruisers like my own *Puffdragon*, ocean-going yachts and fast expensive power boats share the waters with a vast array of commercial craft. A world where only the ducks raise their voices in protest.

Over the years I have cruised most of the waterways of the Netherlands and I have written this book with the cruising family in mind, trying to make it applicable whatever type of craft they may be using. It is intended to be a help in planning the cruise and a reference book to keep aboard during the voyage. It is written to be specific to the waterways of the Netherlands and does not contain general material that has already been admirably covered in other books in this series, particularly *Holiday Cruising on Inland Waterways.*

Experience has taught me that the Dutch produce the best charts of their own waterways and I recommend their use. The Dutch spellings of place names are used throughout. Thus Rhine becomes Rijn; The Hague, 's-Gravenhage; Flushing, Vlissingen and the Dutch word *Kanaal* is used where it refers to the name of a special canal. The more general Dutch term *Jachthaven* is used in this book to cover any type of yacht harbour from a fully fledged marina down to one providing little more than a shore based 'loo', water and a place to tie up.

In the text *yacht* is used to cover both motor and sailing yachts. Where it has been necessary to differentiate, *dinghy* is used to denote a small sailing boat with centre-board and no ballast; *motor boat* to mean a trailerable motor craft regardless of whether it has sleeping accommodation; *motor cruiser* to imply larger power driven craft; and *sailing yacht* to mean any form of sailing cruiser with sleeping accommodation and having an overall length of, say, 20ft or more.

Good cruising to you and if you see *Puffdragon* when you are in the Netherlands, drop in for a yarn.

Springhead School
Gravesend
Kent

J. P. Oliver

CHAPTER ONE

Why Holland?

Although Holland is a highly industrialised country, the industry is largely confined to a small area in the province of Zuid Holland. The rest of the Netherlands is rural country, with small towns and villages almost totally devoid of urban sprawl. The countryside is not so much unspoiled as lovingly and painstakingly created and passionately preserved. The Dutch have fought for their land, not only against human aggression but against their ever present enemy, the sea.

Over most of Holland, good canals and navigable rivers lead through rural scenery and into the hearts of small picturesque towns. It is as well, however, not to spend too much time gazing at the countryside, for the same canal that you are using is being used by barges of 2,000 tons and upwards. Not only are the waterways of Holland extensive, they are very varied. The small car-top dinghy and the large ocean-going yacht alike can find water suitable for their particular enjoyment, though these extremes will be rather more limited in their choice of cruising grounds than will other boats. The cruiser built for comfort rather than speed, and the speedboat and water-skier, will find Holland has plenty of facilities to suit them. In fact the waterways of Holland can cope with all sorts of water-sports enthusiasts, but I must emphasise that most of the rivers, canals and lakes are used by commercial craft, and often large commercial craft, to an extent undreamed of by people used only to English inland waterways.

In parts of Holland, on the busier routes, the locks are enormous and the barges gigantic, going as far as Switzerland at a fair rate of knots. But equally there are lakes and canals where my 26-footer becomes a towering giant and the ducks, disturbed for the first time in days, swim off in a chorus of angry quacking. Over all these diverse waters there persists an air of quiet efficiency and tolerance. Apart from the ducks, each user of the waterways respects the rights of others to use the waters, and all behave in a way to cause least inconvenience to other people.

The waterways of Holland abound in small jachthavens where it is safe to tie up for the night. These range from full marinas with every facility, through town harbours (often old fishing ports) to a few stakes driven into the lake bed with perhaps a small jetty. They are normally in good repair and the charges are largely in proportion to the facilities. All locks and bridges in Holland are controlled by paid officials. You must not attempt to work them yourselves. Apart from a few of the busier commercial canals the locks and bridges operate from about dawn to dusk. On some there is a small charge and you may have to give your name, the name of your ship and a few other particulars. The charge is rarely more than the equivalent of five pence.

The Dutch are a civilised and tolerant nation who have given refuge to many peoples driven out of their native lands by persecution throughout the ages, and they are therefore used to foreigners. During the past 2,000 years they have fought for and won their country from the sea. They have also had to cope with their share of human aggression. The Vikings, the Romans, the Franks, the Spaniards, the French, the Germans and, if my history is correct, there was a certain contretemps with the English. From this the Dutch have emerged as a prosperous, hard-working nation with a strong maritime tradition and a knowledge of hydraulic engineering second to none. The waterways system of the Netherlands is in excellent shape, for the canals not only provide a watersports Utopia, but also serve as the arteries along which the bulk transport of the country flows, and are an integral part of a complex system that keeps a country largely below sea level from drowning.

Dutch history and geography provide the holiday cruiser with two bonuses. Firstly, the visitor is in no danger in the countryside or in the streets of a Dutch town at night, for the Dutch have too much to do coping with the violence of the waves to tolerate a wave of violence in their land. Secondly, a nation that has created over half its own land surface does not go in for vandalism.

CHAPTER TWO

Preparations

The precise details of the preparations you would have to make before cruising in the Netherlands will depend largely on the type of cruise you have in mind, the duration of your proposed stay, your experience and the time of year. The first choice that has to be made is whether you are going to sail across the North Sea in the boat you intend to use for your cruise; whether you intend to take a boat from England trailed behind a car on the ferry; or whether you intend to hire a boat in the Netherlands.

SAILING ACROSS

Firstly you will need to know that your boat is seaworthy and that you have enough navigational and off-shore experience. Crossing from the south-east or east coast of England will present no difficulty for a reasonably experienced off-shore cruising man. If you are taking your boat anywhere outside the UK it is necessary to inform Customs of your departure and to report to Customs on your return. You should not land until the Customs officer has visited your ship and told you that you may do so.

You should carry a spares kit for your boat, particularly engine spares. In general most ships chandlery is easily available in Holland. Dutch marine engineers and sailmakers are highly skilled and most obliging. If your boat is fitted with an English engine, spares may well not be available, and even a $\frac{3}{8}$in Whitworth nut may present a problem in a country that has been metricated for many years.

A seagoing boat may well have a marine toilet. At present these are not sealed in Holland. No one will object either to you emptying a few tea leaves or other vegetable waste over the side, but plastics, ropes, and other non-biodegradable materials that foul propellers and choke water inlets should be kept on board and disposed of ashore. Oil and chemical waste should never be put into the water as they create very great pollution problems. Camping Gaz is obtainable in Holland, but not Calor gas, so that if you cook aboard by the latter it is as well to buy an adaptor before you leave. Likewise for those who shave, shaving points are available in most jachthavens, but the Dutch use a two-pin plug of different pitch to the English plug and it is no use thinking that you can plug into the light socket since the Dutch use screw-in bulbs. The solution is to carry with you, or buy in Holland, a shaver supply lead with a continental plug.

Only a limited number of waterways in Holland are suitable for use by seagoing yachts with fixed masts. The maps on pp 20–1 show those waterways that are navigable by such craft. Reliable auxiliary power is imperative. Power boats have less restrictions, but it should be emphasised that on many of the waterways there are speed restrictions. When crossing the sea, motor cruiser skippers should realise that they are far less visible than a sailing boat of equal length. An efficient radar reflector mounted as high as possible is a most valuable aid for small motor cruisers.

TRAILED BOATS

Taking a boat to Holland by car and ferry has three main advantages over other methods of cruising in the Netherlands. Firstly, although the boat would have to be sound and lake and estuary worthy, it need not be as big and therefore not as expensive as a seagoing vessel. Secondly, no off-shore experience is required by either the skipper or the crew. Thirdly, since Holland is criss-crossed with excellent motor roads, you can move with your craft from one area to another very fast and easily.

Many jachthavens have facilities for slipping such boats and the charges for using these facilities and for parking your car while you are away are very reasonable. It is also possible to slip small boats at many of the ferry crossings on the inland waters by arrangement with the ferry officials. People cruising in trailed boats should note that the Dutch inland waterways system intersperses canals with lakes and therefore such boats should carry lifesaving equipment up to the standard for an off-shore cruise around the British Isles. *How Safe is your Craft?* sets out in detail recommended safety equipment for small craft under these conditions.

The car ferries usually charge for trailed boats as for caravans, so that provided the trailer and car do not exceed the maximum permitted length, separate charges will be made for the boat and the car based on the length of each. The design of ships is constantly changing, therefore if you are thinking of towing a boat to Holland it would be as well to check on the up-to-date regulations before making any detailed plans.

Water-ski enthusiasts should note that there are areas set aside in the waterways where it is safe and permitted to ski. Details can be obtained from the Netherlands National Tourist Office in London or from the *Nederlandse Waterski Bond.*

Trailed boats are not subject to import restrictions in the Netherlands provided that they are being imported for personal use for a limited period. It is however necessary to obtain a certificate of ownership from the Royal Yachting Association. You should also inform your insurance company of your intended visit before you go as they will probably wish to adjust your premium.

At the small end of the trailed boat scale comes the trailed

Page 17
(*above*) A typical windmill at sunset
(*left*) Late-Gothic town hall at Gouda

Page 18 (*above*) The magnificent Zeelandbrug; (*below*) storm gates at Krimpen a/d Ijssel

dinghy. Young people in Holland do a vast amount of dinghy cruising. The basic requirements for this type of cruise are youth, a good companion, a dinghy between 14ft and 17ft overall (not a hairy racing plank but say a GP 14, an Enterprise, a Wayfarer or a Mirror 16), some basic camping gear that can be stowed on board and a good pair of paddles or a small outboard motor. You should take a red ensign and fly it as this will amuse the Dutch dinghy sailors and help break the ice. Then do as the Dutch do. Sail when you like. Otherwise pull into the side of the lake or canal, put up your tent cover or a tent, and either stay there or go on the next day. You are free to go more or less where you wish. You will not be regarded as a poor relation or a scrounger, but will just be accepted as another legitimate user of the waters who is doing his own thing. The best area for this type of cruising is Friesland, although any of the lakes are suitable. Incidentally, if you run out of wind towards evening and don't feel like rowing, hail the nearest passing cruiser and you are almost certain to be taken in tow. Keep clear, however, of the busier commercial canals except at weekends.

HIRING A BOAT IN HOLLAND

There are many boat hire firms in Holland and the industry is expanding every year, but the demand far exceeds the supply. It is thus essential to make early application and to try several firms. A list of boat hire firms can be obtained from the Netherlands National Tourist Office in London. The range of craft available is considerable and varies from traditional (reproduction) Dutch sailing boats, not for the novice unless he wants to find out how to sail the hard way, but an experience for the competent sailor of modern yachts, to modern family sailing cruisers normally of Dutch design but varying little from their English counterparts. There is also a vast array of motor cruisers.

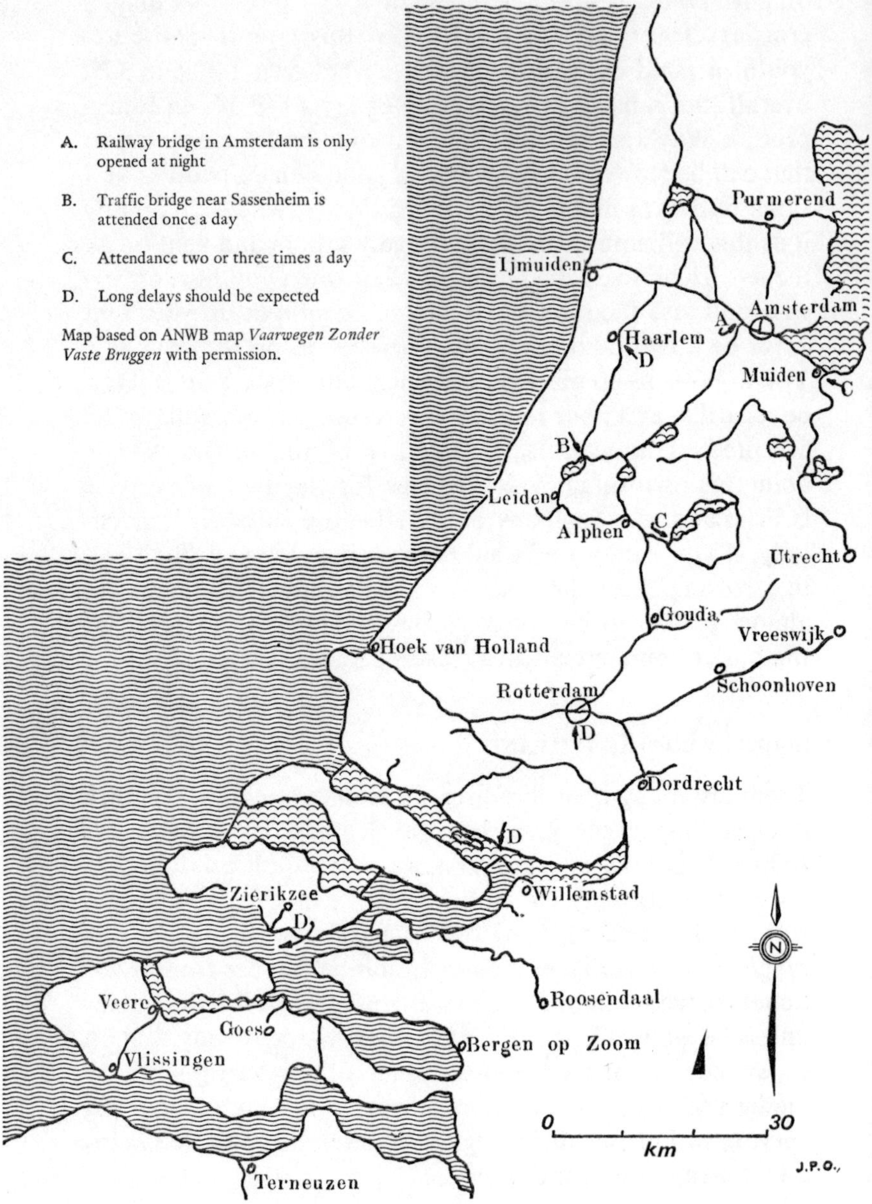

Waterways navigable by yachts with fixed masts (southern part)

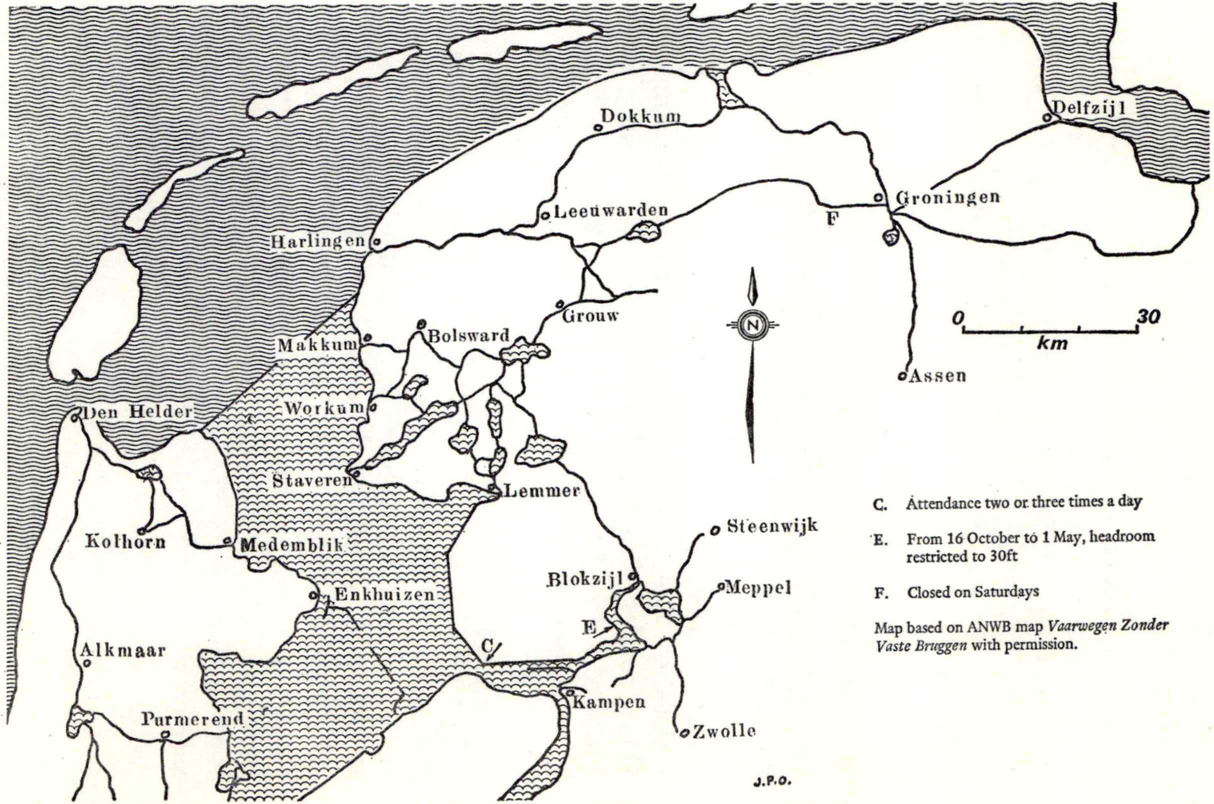

Waterways navigable by yachts with fixed masts (northern part)

OTHER METHODS

If you have a boat in England, you can advertise in the Dutch or the English yachting press which is widely read in Holland, that you would be prepared to swap boats for a period with the owner of a similar boat in the Netherlands. If you prefer to be driven during your cruise, a number of 'Boatel' cruises are available.

FORMALITIES

The Netherlands Government has a very kindly attitude to amateur sailors, in particular to those from other countries; therefore formalities are minimal. A visiting yacht, be it trailed or sailed from England, must wear the flag of its nationality. You should fly from the starboard yardarm a Dutch courtesy flag (and I would advise those sailing in Friesland to fly a Friesian courtesy flag as well). If your boat is not registered you should have a certificate of ownership obtainable from the RYA. You need no licence nor a certificate of competence.

You must make sure that your boat is insured for the waters you are to sail. If you wish to be within the letter of the law you must have on board Part 1 of the *Almanak voor Watertoerisme*, which is published by the ANWB and obtainable from them. The Almanak contains all the rules about Dutch waterways needed for pleasure craft. It is however written in Dutch.

All on board must have valid passports but UK citizens need no visas. There are no other formalities.

CHAPTER THREE

The Cruising Grounds

For the sake of convenience I have divided the waterways of Holland into six cruising grounds. Although their boundaries are geographical I have drawn arbitrary lines so that a particular area includes a type of cruising ground more suited to a particular type of craft. The waterways system is, however, so complex and varied that no rigid demarcations can be made. All the waterways are interconnected and interrelated and so I have included a few through routes to help you to choose the best route from one area to another, and a few that for one reason or another I would not recommend, though looking at the charts they may seem logical ways to go. One factor to be considered in making a decision where to sail is the type of boat in which you will be cruising. Another factor, for those sailing from England, is the time that you will have to spend getting to your cruising ground and back to England after your cruise. The time involved will depend on the location of your home port and you will have to allow for possible delays on the return voyage due to bad weather. At the end of some seasons the yacht harbour at Vlissingen is full of British boats whose owners have had to leave them behind and return to England on the ferry because the weather has been too bad to sail. For many, particularly those using their own yachts, the southwest of the country is probably the most likely area, and indeed, this is where most British cruisers are to be found.

Although the Netherlands provides cruising water for

almost any type of boat, the most suitable craft for sailing across the North Sea and then using for cruising would have the following specification:

Length over all	23 to 35ft
Draught	Not exceeding 3ft 6in
Height (above waterline)	Capable of being reduced to 6ft
Speed	For a sailing boat 5 knots under power
	For a motor cruiser, to be able to cruise at 5 knots
Beam	Not more than 9ft (catamarans are rare in the Netherlands)

Such a boat could cruise most of the waterways, is big enough for the large lakes yet small enough to penetrate the smaller and delightful little towns to be found along their banks. Sailing boats with fixed masts will only be able to sail in the Schelde and the Veersemeer, on the Ijsselmeer, in parts of Friesland, the Waddenzee and a few of the canals.

THE SCHELDE AND THE VEERSEMEER

This vast delta area provides scope for cruising for all types of boats and all levels of experience. There are few locks and no fixed bridges. Canals are few and short.

Vlissingen and the Kanaal door Walcheren

Since for many British boats this will be their first experience of Dutch waters, and for many off-shore cruisers their first experience of locks and inland waterways I will start at Vlissingen and take you through the canal to the Veersemeer. Vlissingen is at the south-west extremity of the island of Walcheren and to lock into the Kanaal door Walcheren a boat has to enter the Buitenhaven which is about one mile to the east of the Vlissingen lighthouse. The entrance, marked by a fixed red light to be kept to port and a fixed green light to starboard, is easy to find since the harbour is used by the

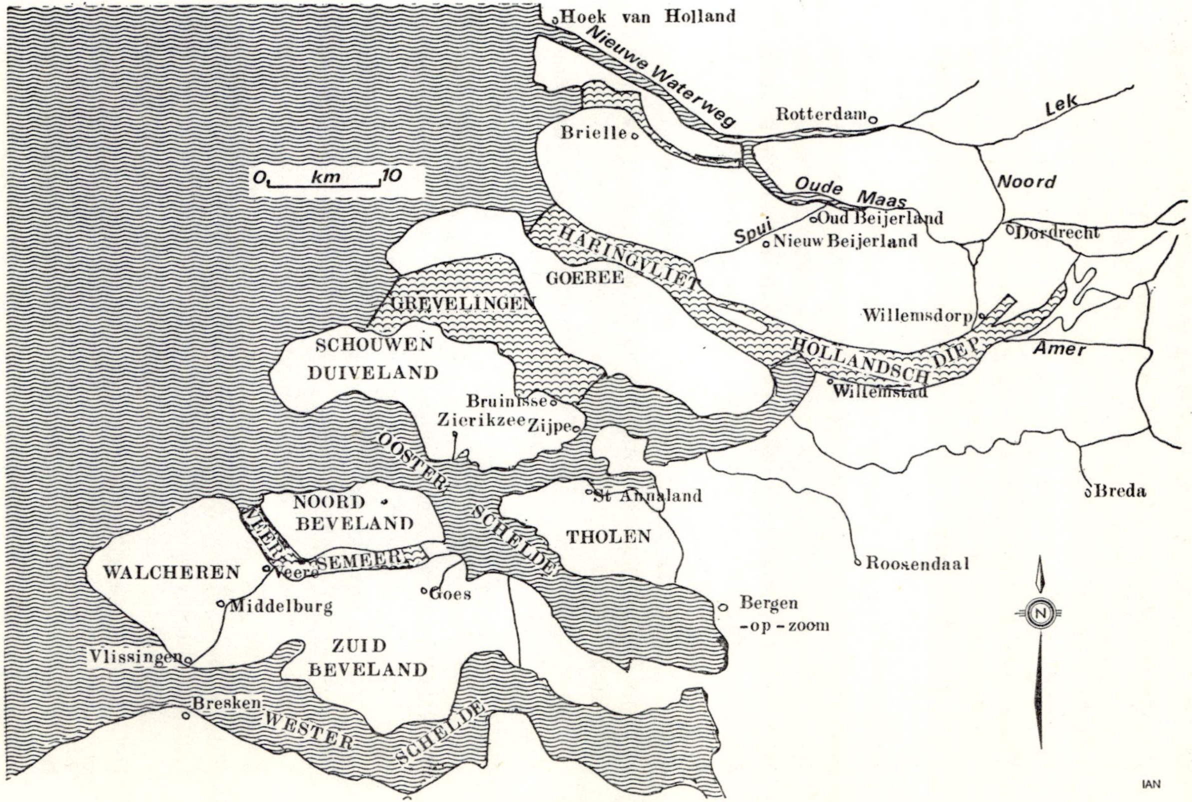

Waterways of southern Holland

regular ferry service to Breskens on the southern bank of the Westerschelde. Sail round the breakwater, taking care not to have an altercation with the ferry, and the lock gates will be seen on the port side, just to the left of the ferry terminal. Traffic through the locks is controlled by light signals (see Chapter Five).

If you have to wait it is customary to tie up to the piles or another waiting boat on the port side of the harbour. When the lock gates open and you get the green light entry signal, first allow any commercial craft to enter and tie up. It is important to let them tie up as these barges manoeuvre in confined spaces using large rudders to deflect the wash from very powerful engines. Fighting the wash from these propellers with your light displacement craft may be exciting but is very bad seamanship and can be fatal. Enter slowly and tie up. You have nothing else to do. The locks will be operated for you. Unless otherwise instructed, let the commercial craft leave first when the gates open.

You are now in the canal. On the right are two commercial harbours and after passing these the canal bends to the right. Just before the first bridge you will find on the right the jachthaven, where you can obtain a quiet mooring, a shower, beer from the fridge in the club-house and diesel fuel for the boat. Petrol can be obtained from a garage not far away. There is always a warm welcome here for British boats and the harbourmaster is a mine of information and a wizard at weather forecasting, a piece of information to be kept in mind for the return trip. The water here is always warm, being on the outflow side of the cooling water from the local power-station. This jachthaven has a slip.

All the bridges on the Kanaal door Walcheren open during the day but not at night. Half-way between Vlissingen and Veere is the charming town of Middelburg where you may tie up along the canal bank, or after passing through the two bridges you can turn left into the Voorhaven and then sharp left again into the jachthaven. Middelburg, the capital of

Zeeland, is a good shopping and sightseeing centre. As with all Dutch towns you should contact the local VVV office for information about what is going on and what to see. The market at Middelburg held every Thursday is a tourist attraction as well as being a working market. Many of the stallholders and local people dress in their provincial costume for this weekly event. Although the town suffered a great deal of damage during the 1939–45 war, all has now been restored and there are many buildings of historical interest. The 'Miniature Walcheren' exhibition and the *son et lumière* given in the town during the summer months are well worth visiting. Another four miles up the canal, and you have to lock out at Veere into the Veersemeer.

Veere and the Veersemeer

Veere is a town of great charm and well worth visiting. Once a town of over 10,000 inhabitants, Veere has declined in importance and now has a population of less than 4,000. The fifteenth-century church and the town hall which now houses a museum are of particular interest. The sixteenth-century 'Scottish Houses' are reminders of the days when Veere was an important port and had a thriving trade with the Scots. The Veersemeer is a quiet lake sandwiched between the once separate islands of Walcheren, Noord Beveland and Zuid Beveland. Once part of the estuary but dammed off in 1961, it is now only possible to enter it through either the Kanaal door Walcheren or the lock in the Zandkreekdam from the Oosterschelde. Commercial traffic is very light and the meer is well provided with jachthavens and is suitable for all types of cruising boats. Hire cruisers are also available in this area.

The water of the Veersemeer is extremely clear and the bottom is sandy. The area is rural and the shores contain two nature reserves. There are a number of small islands in the meer, many of which have piles or small jetties for yachts to tie up as well as several fully-fledged marinas. The meer is used by the Dutch very much as a water-sports holiday area, but more in the getting away from it all style, than as an area of bright lights and dancing till the early hours. Sheltered and safe, with interesting towns, often containing reasonable restaurants, and with mooring places not far from each other, the Veersemeer provides an ideal family cruising ground for larger boats, trailed boats and dinghies.

The Schelde and waterways to Rotterdam

In the Schelde estuaries the cruiser has to cope with relatively open water, strong tides, shifting sandbanks and fair distances between safe harbours. Delightful cruising for seagoing boats, particularly for sailing yachts that can really spread their wings and for the faster motor cruisers. I include in this area the Nieuwe Maas from the Hoek van Holland to Rotterdam in the north, the Oude Maas leading to Dordrecht, the Haringvliet, Hollandsch Diep, Grevelingen, Volkerak, the Oosterschelde and the Westerschelde. Although divided now into two halves by the locks at Willemstad, the whole area forms the same type of cruising ground with the same problems and delights.

The Westerschelde

At the extreme south of the area, the Westerschelde leads from Vlissingen eventually to Antwerp and is the only route for seagoing ships to the Belgian port. The tides are strong and the sandbanks constantly shifting. Before sailing these waters it is essential to obtain a detailed and up-to-date chart, preferably a Dutch one. Last year's chart will definitely not do. A seagoing boat and experience are also essential. Apart from Vlissingen, the only other Dutch harbours suitable for yachts on the Westerschelde are: Breskens, a holiday resort and ferry terminal; and Terneuzen, once a fortified town but now a growing industrial one. Breskens is on the south bank opposite Vlissingen. The jachthaven is about half a mile east of the ferry harbour. The entrance leads to two havens, and the jachthaven is the eastern one. Terneuzen also has a jachthaven, situated to the east of the entrance to the much larger Buitenhaven that is also the entrance to the Ghentkanaal. This canal is very busy and entering the jachthaven necessitates crossing the main stream of barge traffic. Yachts should *not* tie up at Hansweert on the northern bank, nor use the Kanaal door Zuid Beveland.

The Oosterschelde

The Oosterschelde has a rather different character from the Westerschelde, mainly because fewer commercial seagoing craft use it. Entering the Oosterschelde from the sea will soon be impossible, as this is one of the arms of the estuary to be closed. At present, except in excellent weather, it is highly undesirable to try to negotiate the sandbanks at the entrance and there are many wrecks in the area to prove the point. Yachts, however, enter the Oosterschelde from the Veersemeer after a mile of very narrow channel. This Zandkreek channel leads from the locks in the Zandkreekdam on the Veersemeer into the large expanse of water and sandbanks that is the Oosterschelde. About one mile to the east of this channel is the entrance to the canal leading from Sas van Goes to the town and the minute but charming jachthaven at Goes. A visit amply repays the small charge for using the canal. The central square of the town contains most of the historic buildings and for those who love a good story that of Jacoba, who lived here in the fifteenth century, is one of Holland's greatest true life romances. Farther south-east leads you into the main stream of barge traffic and after this you become entangled with the works associated with the building of the new canal that is to connect Antwerp with Rotterdam via Moerdijk and Dordrecht. It is therefore advisable *not* to proceed eastwards of Sas van Goes, nor to visit Bergen-op-Zoom except by road nor at present to use the Eendracht which runs to the east of the island of Tholen.

Leaving the Zandkreek channel and turning north-west you soon see the magnificent Zeelandbrug, with a height of about

35ft under the spans. This impressive bridge, opened by Queen Juliana in 1965, is the longest of its type in Europe. It is built of prefabricated concrete sections on 162 caissons. About a mile to seaward of the bridge are two harbours suitable for yachts. On the north shore is Zierikzee, reached up a short canal. Zierikzee is a very ancient town partly surrounded by canals and accessible by road through picturesque old gateways and quaint bridges. The St Lievens Monster Tower dating from the fifteenth century can be climbed and from the top you get a magnificent view of the countryside and estuaries. Yachts entering Zierikzee should moor as directed by the harbourmaster who seems to spend a great deal of energy cycling around the pontoons directing his flotilla. Visiting yachts are made most welcome, and the facilities seem to grow annually. There is fresh water, fuel and a ships' chandlers on the quayside, and showers that require a 50 cent piece to operate them. There is, however, no Dutch 50 cent piece, so you have to go to the address, just round the corner, given on the door of the ablution block and buy a token.

Colijnplaat on the southern bank is primarily a fishing harbour but there are facilities for yachts to tie up. I bought the best fish I have ever tasted direct from a trawler in Colijnplaat, but unless you are passionately keen on fresh fish or wish to stand and marvel at the control of the trawler skippers as they bring their boats into this rather small harbour, there is no reason to visit Colijnplaat.

Zierikzee to Willemstad

A channel leads eastwards from the Zeelandbrug and passes between Schouwen Duiveland and Tholen. This provides excellent cruising waters, but there is also a fair measure of barge traffic. There are many small towns that can be reached through small canals. They are accessible for shallow-draught yachts and the entrance channels are normally buoyed or marked with stakes. Most of these towns have a place for yachts to tie up, but few have much more to offer the cruising man. St Annaland, reached up the Krabbenkreek, has a good jachthaven, but at present it is unwise to go farther along the Krabbenkreek than St Annaland, because of the canal works already mentioned.

The main channel narrows and the strength of the tide increases as you pass through the narrows at Zijpe, where the old ferry looked more like a Mississippi steam boat than a Dutch ferry, for it invariably belched forth great clouds of black smoke. Unless you have a powerful boat it is best to plan to pass through these narrows with the tide. The tramweghavens on either side of the channel here are strictly for the ferryboats, but the Vluchthaven on the western shore has facilities at its southern end for yachts to tie up. Yachts should not tie up elsewhere in this harbour as it is primarily intended as a place for barges to rest at night. It is, however, preferable to the harbour at Bruinisse unless you arrive during the day and can pass through the locks at the southern end of the Grevelingendam and into the jachthaven at Bruinisse.

If you do pass through these locks you will be in delightful cruising waters, with very little commercial traffic, in what was until recently the Zeegat van Brouwershaven and open to the North Sea. However, the dam is now complete. The harbours at Brouwershaven, the town in which Jacob Cats (1577–1660) the great Dutch poet-statesman was born, and Ouddorp on the north shore are suitable for yachts. The area is non-tidal but rather shallow.

If instead of passing through the Grevelingendam you sail eastwards through the Krammer and Volkerak, in about twelve miles you will reach the enormous locks at Willemstad. You will still be on a main barge route, and although up to this point there has been ample sea room, all traffic, barges and yachts, has to pass through the same locks.

The Locks at Willemstad

The overriding condition for a safe passage through these

locks is not to panic. Everyone else knows what they are doing. Most of the barges have been through the locks many times before. The whole operation is very carefully controlled and the yachtsman has only to follow a few simple instructions. These are the largest and busiest locks that most cruisers are likely to meet in Holland and except at weekends the commercial traffic is very heavy. Here is the best procedure.

Approach with caution, keeping out of the way of commercial traffic going in and out. Remember your manoeuvreability is greater than theirs. At each end of the lock a wall to tie up to extends for some distance. See which lock is to be used next and, unless you receive directions over the loudspeaker system to the contrary, tie up to the wall at the end farthest away from the lock at the one of the pair of locks that has the gates towards you closed. Wait. If there is a Dutch yacht tied up it is a good thing to tie up near him and engage him in conversation. If the directions are given in Dutch he will probably translate them for you. The commercial traffic is normally called in first and by name. Wait for your yacht's name to be called, or if your boat has a name that is unreadable by the control or unpronounceable in Dutch you will be addressed as *Engels jacht*. Remember that *backboord* is the port side of the lock and *stuurboord* the starboard. In general the control will keep the message simple if it is given in Dutch, such as *Engels jacht, stuurboord.* However the directions for English yachts are normally given in English. Beware of the wash from propellers and tie up exactly as directed. The space may appear to be very small and at first you may not think it is large enough. The general plan seems to be to get as many of the larger barges in as possible, and then to fill the gaps with smaller barges and yachts. Thus a barge may be turned away from one lock while a smaller yacht arriving later may be accommodated.

There are no formalities and no dues to pay. When the lock opens, remain firmly tied up as the wash from the propellers of engines that can drive 2,000-ton barges against a 12-knot current (for many of these barges sail up the Rijn to Switzerland) will overpower anything even a powerful motor cruiser can manage. When all is clear, leave the lock as quickly as possible as there are craft waiting to come the other way. Sailing yachts with small auxiliaries should beware as they leave the locks. Near the open gates there is considerable turbulence caused by water coming in at the side of the locks, so yachts with small engines should keep to the middle. I have passed through these locks many times and the only damage I have ever suffered was from an English motor yacht whose owner prefaced his approach with 'I've just come across the North Sea single-handed. Can I tie up alongside?' I agreed and lost 3ft of rubbing strake. He opened another can of beer and I made a mental note that I wasn't surprised he was single-handed. Passing through the Willemstad locks is like learning to drive a car in London. Once you have mastered it, traffic never worries you again.

The Haringvliet

North of the Willemstad locks you can turn north-west and pass under a bridge which has a height under the spans of about 40ft depending on the state of the tide. There is an opening span for taller sailing yachts with fixed masts. This brings you into the Haringvliet where good cruising water and very light commercial traffic is to be found. This stretch of water is dammed off from the North Sea, but is tidal in that direct access to the North Sea is gained through the Oude Maas and the Nieuwe Maas, and then via the Spui or the Dordtsche Kil. There are harbours for yachts at Middelharnis reached through a short canal for which there is a small charge that seems to include the berthing fee; at Hellevoetsluis on the northern bank; and there are several other small mooring places for yachts. The Spui connects the Haringvliet via Goudswaard, Nieuwe and Oud Beijerland, with the Oude Maas. The Spui is a quiet river and certainly a better way for sailing yachts to reach Rotterdam than the busier route via the Dordtsche Kil and the Noord.

Willemstad and the Hollandsch Diep

Most people leaving the Willemstad locks for the first time feel in need of a little rest and a quiet beer. For these people the route should be eastwards passing the werkhaven, denoted by a fixed green light to the west and a fixed red to the east, and then round the small mole bearing a fixed green light. Remember that the water is still tidal and the tide strong even though you have just passed through a lock, then enter the short channel (about 50yd) before turning left at the signpost into the jachthaven at Willemstad. This jachthaven has all facilities and can cater for all kinds of yachts. During the school holiday period you should watch out for children who often swim in the entrance to the harbour. Willemstad is a very picturesque town and both the town and the jachthaven are dominated by an enormous windmill that is floodlit at night. The jachthaven is often crowded but there always seems to be room for one more. There is a town harbour but it has no facilities. Normal provisions can be bought in the shops at Willemstad but they offer nothing more.

The Hollandsch Diep is a good stretch of cruising water for larger boats, reaching for about six miles to the east of Willemstad and terminating at Willemsdorp. There are mooring places for yachts at Moerdijk on the southern shore and at Strijensas on the north bank. There is a constant stream of barge traffic through the Hollandsch Diep but this divides like the channel at Willemsdorp, part of it going northwards through the Dordtsche Kil towards Dordrecht and Rotterdam and part going under the bridge, which has no lifting span, into the Nieuwe Merwede. There is a jachthaven at Willemsdorp.

Dordrecht to Rotterdam

A the notrthern end of the Dordtsche Kil is the town of

Page 35 (*above*) The author tries his hand at the 'national pastime'; (*left*) one of Holland's rare deep locks

Page 36 (*above*) Waterbus in Amsterdam; (*below*) river police boat at Muiden

Dordrecht, which from the water is largely industrial, and although there are six jachthavens in the town they are difficult to find. If you particularly want to visit Dordrecht then it is best to pass under the two bridges, which have very restricted opening times and so present a barrier for sailing yachts with fixed masts, and find shelter in one of the two jachthavens on the Wantij to the north of the town. These are set in semi-parkland and are both quieter and more easily found than the more central jachthavens and yet are only a mile or so from the centre of Dordrecht. The town itself is partly industrial and partly historic. The VVV can give you all the sightseeing and tourist information you may require. It is worth mentioning that Dordrecht stands at the busiest river junction in western Europe and something like 1,500 commercial craft pass the town every day.

From Dordrecht, Rotterdam can be reached via the Noord, running north from Dordrecht, then along the Nieuwe Maas. This route is largely industrial and carries the bulk of the barge traffic, so that unless you are in a hurry and have a powerful motor cruiser then the alternative route through the Oude Maas is recommended. The Oude Maas is a river of some charm with light commercial traffic, although individual boats can be quite large. It is well supplied with good jachthavens and has only two bridges, both lifting ones, at the western end. It is quite possible for a sailing yacht to tack in the Oude Maas but it would be unwise to do so in the Noord.

Rotterdam

The Oude Maas finishes at its western end in the Nieuwe Waterweg at the eastern extremity of the Europort industrial complex. At the end of the Oude Maas, therefore, you have two alternatives, to head westwards through most of the industrial area and reach the sea at the Hoek van Holland or to turn east and enter Rotterdam. Although Rotterdam is worth a visit, the shopping centre is excellent and is in pedestrian precincts, it is more a place to stop at when going from one place to another, for this is not cruising country.

There is only one jachthaven at Rotterdam, the KRZV de Maas (Royal Maas Yacht Club). This jachthaven is not difficult to find and is in the veerhaven, the first harbour on the north bank after passing the Euromast. Visiting yachts are made most welcome and the harbourmaster is always very obliging. Although the Maas is very chopped up the harbour is remarkably quiet. The gate from the street to the jachthaven is locked at night so visitors should obtain a key, against a small deposit, from the harbourmaster if they are expecting to return from town after nine in the evening. Rotterdam has everything that could be required by a visiting yacht and the harbourmaster can supply information about where to find what you want. For more general information you should take a short tram-ride into town and visit the VVV office.

The Rotterdamsche Waterweg (*Nieuwe Waterweg*)

From Rotterdam you can reach the sea directly down the Rotterdamsche Waterweg. This important waterway carries a large amount of ocean-going shipping and flows through the whole length of the Europort industrial area. On Sundays it is relatively quiet but on weekdays very busy. At any time of the week sailing along it is quite an experience. There is only one suitable mooring for yachts between Rotterdam and the Hoek van Holland, in the Binnenhaven at Maassluis, about half-way down the waterway. Yachts have no official place to tie up at the Hoek van Holland, but if they wish to stay for a day or a night in order to catch a suitable tide or to rest after a North Sea crossing, they should go to Berghaven on the north bank just to seaward of the Harwich ferry terminal. Berghaven is the pilot boat harbour and yachts, although not officially welcome, are not turned away. For skippers proceeding seawards from here it is essential to obtain the most detailed and recent charts, as work is in progress on the entrance to Europort. It must be emphasised that the Rotterdamsche Waterweg is suitable only for sea-going boats and provides a cruising experience for experienced cruisers not obtainable anywhere else in the world. Sailing yachts may sail but should not tack.

The Hartelkanaal and Brielsemeer

South of the Rotterdamsche Waterweg and running parallel to it is the Hartelkanaal, leading off from the Oude Maas. *The Hartelkanaal is forbidden to yachts.* Just south of it, however, is the Brielsmeer, a small non-tidal water reached through the Voornse sluis from the Oude Maas. This water is used by yachts and has several good jachthavens. It is dammed off from the North Sea and so leads nowhere except to the town of Brielle, the Brill of history books. It is good cruising water and much used by Dutch sailors from Rotterdam, but unless you are very lucky with the wind, the smell from the factories and refineries just to the north can be offensive.

THE IJSSELMEER

Before I set sail for the first time on the Ijsselmeer an experienced Dutch sailor warned me that it was the most dangerous piece of water in Europe. I have now sailed across this inland sea many times and would endorse his warning, but I would also add my unqualified opinion that it is the most rewarding single lake for the cruising man. I have been on the Ijsselmeer in all weathers from thick fog to brilliant sunshine, and from a flat calm to a raging gale during which I sailed under bare poles. This water is only dangerous because, being a lake, many visitors underestimate its power. No Dutchman ever does. As if to emphasise the point the Dutch maintain lifeboats, radio beacons and lighthouses there in the same way as in open waters around the coast. With a seaworthy craft and a healthy respect for this lake this one-time Zuyderzee is well worth a visit, but you will have to go soon, for the reclamation works now in progress may well change the whole character of the area.

The lake is not very deep, on average only about 12ft, and because of this a strong wind can, in an incredibly short space of time, build up short steep seas that will stop a beamy bluff-bowed boat dead, or provide an exceptionally wet ride for those in sleeker craft. The waves are just as quick to die down when the wind has passed and in an hour or so the surface can again resemble a mill pond. All around the lake are excellent harbours left over from the days when large fishing and trading fleets sailed directly into the North Sea. The towns are both picturesque and full of history. Apart from these older harbours there are many new jachthavens with every possible facility to cater for all tastes. These, together with the low density of commercial traffic, which is confined mainly to a route from Amsterdam along the southern and eastern shores to Lemmer, makes the Ijsselmeer a family cruising paradise.

From Enkhuizen in the west to Lelystad in the south, a dyke is being constructed and the area south and west of this dyke will be pumped dry with the exception of one or two canals. Anyone used to the more leisurely pace England adopts to works on inland waterways will be surprised by the speed with which such works are carried out in Holland. This is not said to decry the English efforts but to emphasise that cruising on the Ijsselmeer with last season's chart is at best inconvenient and at worst downright dangerous. The Dutch small boat charts are reprinted every April and anyone cruising this area should obtain the latest.

To the south of the Ijsselmeer lies a much smaller section of far more sheltered water lying between the reclaimed Zuidelijk and Oostelijk Flevoland and the provinces of Utrecht and Gelderland. Smaller craft will be more at home on this water with its little islands and small towns and harbours on the southern and older bank. Small craft in good weather can sail on the Ijsselmeer, but would be well advised

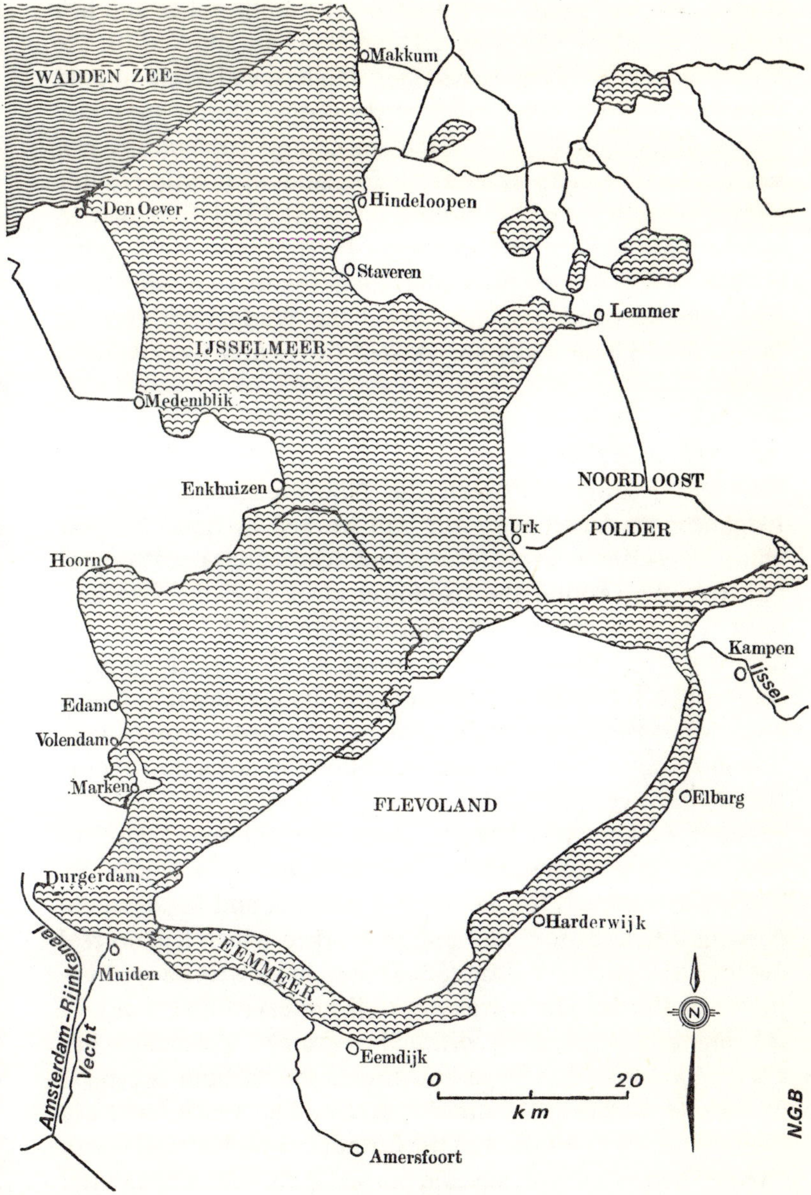

The Ijsselmeer

not to go out in unsettled conditions. There are no speed restrictions, but there are a number of fishing stakes and fish traps in the shallower water. These areas are marked on the Dutch charts as *Vele visstokken.* High speed craft should therefore only use their power in the deeper clearer stretches. Large sailing yachts with fixed masts and catamarans can sail freely. They can enter from England with little trouble either by sailing up the Noordzeekanaal from Ijmuiden to Amsterdam and locking in to the Ijsselmeer at the Oranjesluizen, or by entering through the locks in the enclosing dyke that separates the Ijsselmeer from the Waddenzee and the North Sea. These locks are at Den Oever and Kornwerderzand.

Amsterdam to Edam

The Oranjesluizen between the Noordzeekanaal and the Ijsselmeer are large, busy and still worked by hand, but the whole operation is very efficient and delays are rare. Proceeding from here round the Ijsselmeer clockwise, the first harbour is at Durgerdam. The entrance channel has a nominal depth of 1.8m though with a strong south-westerly wind it can be less. It is thus suitable only for shallow-draught boats, and indeed the harbour has little to offer apart from a very safe mooring and a small and exceptionally noisy do-it-yourself chain ferry.

North of here you have to round the one-time island of Marken, now connected to the mainland by a causeway. There are harbours for yachts at Volendam and Monnickendam on the mainland, and one at Marken which has no real facilities for yachts. All three harbours have well buoyed entrance channels. The towns are well preserved and many of the inhabitants wear traditional costume. Volendam has many characteristic village houses and the harbour is typical of the old harbours of the Zuyderzee. The inhabitants are predominantly Catholic and the Catholic church is well worth a visit. There is a small museum housed in the Hotel Spaander.

Monnickendam has an interesting church and the Dutch flair for fancy clocks shows in the chiming clock in the church tower where two horsemen appear at the full hour chimes. There is a tile museum on the upper floor of the weigh-house. Marken is noted for its traditional costumes and the fact that it is built on eleven mounds. The dyke connecting Marken to the mainland was completed in 1957. The old Marken lighthouse still stands and is a valuable navigation asset.

The three towns are well worth seeing if you like that sort of thing. For me, they are far too tourist conscious, being visited by hordes of sightseers from Amsterdam during the summer months. The same goes for Edam, famous for its cheese. A little farther north, it has to be reached up a short canal. The town boasts the oldest chimes in the Netherlands. There is also a cheese market and a museum.

Hoorn

Three other important harbours on the western bank of the Ijsselmeer are, I have to admit, more to my liking. The southernmost is Hoorn. There are two jachthavens there, a small one tucked away in the trees in the Vluchthaven and a large one in the Grashaven. I have stayed at both and would suggest that the larger is perhaps better endowed with facilities whereas the smaller preserves a more homely atmosphere. Hoorn is an historic town, with good shops and a more discreet approach to the tourist than its southern neighbours. There are many sixteenth- and seventeenth-century buildings; the Statencollege is now the West Friesian Museum. The old warehouses of the Dutch East India Company still stand, and from June to August there is a street market every Wednesday.

Enkhuizen

To reach Enkhuizen from the south you have to pass through locks in the new dyke, but the traffic is largely pleasure craft.

There is a town harbour, but yachts are strongly advised to use the jachthaven in the shadow of the Zuyderzee Museum. This is large and has excellent facilities. Like Hoorn, Enkhuizen was a Zuyderzee port of some importance but preserves its seventeenth-century characteristics rather better than does Hoorn. The *Drommedaris*, a sixteenth-century town gate, dominates the entrance to the old harbour. The town was once an important herring fishing port and later had a profitable trade in cloves. Now it is the centre of a large flower and seed cultivation area. The Westerkerk was built by rich farmers in the fifteenth century and has a separate bell tower. Not to be outdone, the fishermen, who were as poor as the farmers were rich, built the Zuiderkerk and gave it the highest tower in the town. There is a rare and valuable collection of books housed in the Westerkerk.

The shops at Enkhuizen are good and the whole atmosphere of the town delightful. The Zuyderzee Museum has no equal of its kind anywhere in the world. If you want to know what the Ijsselmeer was like before the enclosing dyke was built, how the people lived and what the plans are for the future, or if you simply want to gaze at real traditional Dutch craft rather than at reproductions of them, then this museum is an absolute must.

Medemblik

Farther north and thus farther off the tourist track is Medemblik. The jachthaven there is reached through a lock. Again facilities and welcome are all that could be desired, but although the shops are adequate for most cruising folk I suggest that it is less of a shopping centre than Enkhuizen or Hoorn. Medemblik is a very old place that was granted town status in 1289, and was a flourishing trading town in the sixteenth century. It declined, however, and only recently started to expand again as a result of tourism and as a watersports centre. To the south of the harbour is the Radboud Castle where exhibitions and *son et lumière* shows are often staged.

The castle was built by Flores V in the thirteenth century and has recently been restored. It was supposedly built on the site where the most famous Friesian king Radboud lived, although the historical evidence for this is extremely thin.

The Afluitdijk

At Den Oever in the extreme north-west of the Ijsselmeer there are mooring places for yachts on either side of the Stevinsluis leading out to the Waddenzee, but no other facilities. Stevinsluis is the lock at the western end of the great enclosing dyke, the Afluitdijk. At this end of the dyke are also three groups of five drainage sluices that control the level of water in the Ijsselmeer. There are other locks at the eastern end of the dyke, but again, apart from short-stay moorings and a passage to the Waddenzee there is nothing for cruising yachts.

The dyke, built in the open sea, was completed in 1932, 18 miles long and over 20ft above sea-level. It is constructed of rock, loam, sand and clay and carries an important motor road. In the middle of the dyke is a café with a small landing stage on the Ijsselmeer side. Above the café is a monument which can be climbed, and from the top a magnificent panoramic view of the Waddenzee and the Ijsselmeer can be obtained. An embossed plaque reads *Een volk dat leeft, bouwt ann zijn toekomst* (A nation that is alive, builds for its future), a sentiment that the Dutch seem to have taken very much to heart. At Den Oever there is a statue to Dr C. Lely, the great hydraulic engineer and designer of the Zuyderzee works.

The Friesland shore

The north-east shore of the Ijsselmeer is in the province of Friesland. I shall have more to say about Friesland in a later section, but I will deal with the harbours that lead on to the Ijsselmeer at this stage. The most northerly is at the quaint town of Makkum, reached from the Ijsselmeer through a narrow channel between two islands. There are mooring places for yachts on the Ijsselmeer side of the locks that lead to the canal that runs through the town, but it is advisable for anyone staying to pass through the locks and tie up against the canal wall in the town. The canal is small by Dutch standards, and almost totally devoid of commercial traffic. The locks and swing bridge in the town are hand operated and even the townsfolk, who must have seen them worked many times, still seem to find it fascinating. The residents of Makkum claim that the locally made blue pottery is superior to Delftware. There is a pottery museum in the town so you may judge for yourself. The local potters were once forced by the occupying French forces to make cooking pots for the French troops; they did as they were ordered but when the pots were delivered they were found to resemble chamber pots.

The canal system of Friesland can also be entered at Workum a little to the south of Makkum, and at Hindeloopen where there is a jachthaven. On the waterfront at Hindeloopen are the 'Summer Houses' where the wives and children of the seafaring men of the town used to live while their menfolk were away at sea. When the family was again reunited they returned to their proper homes in the town.

Farther south still lies Staveren. Yachts should not enter the Spoorweghaven, which is reserved for the ferry boats, but the Nieuwe Voorhaven that leads through very modern locks to the canal. About 10,000 craft pass annually through these locks on their way to the Friesian lakes. Apart from short-stay moorings there are no facilities for yachts at Staveren, but it is a useful gateway to Friesland.

Lemmer is probably the most popular way to enter Friesland by boat from the Ijsselmeer. Facilities for yachts at Lemmer are extensive and excellent, though some parts of the jachthavens are more suitable for shallow-draught yachts. Yachts must enter through the town and the town locks: those to the west of the town leading directly into the Prinses Margrietkanaal should be left to commercial traffic. The town of Lemmer is thriving, with good shops and a spread of holiday houses on the outskirts. It is possible to lie alongside the town wall, but most visiting yachts prefer to use the jachthavens.

Urk and the Noordoost polder

South of Lemmer lies the Noordoost polder, at the southern extremity of which is the town of Urk. Until 1942 Urk was an island in the Ijsselmeer, but when the Noordoost polder was reclaimed Urk was joined to the mainland. The road to Urk was completed in 1948 but in 1951 visiting cars were banned from the town. Urk has a large fishing fleet, and the fish market still operates despite the fact that the fleet now operates from Delfzijl, the fishermen of Urk sending their catches by road to the fish market in their home town. Not so long ago the staunchly religious (Protestant) and isolationist inhabitants of Urk would, by all accounts, have stoned any yacht daring to visit the town on Sunday. However, times have changed and so have habits, and it is possible to spend a quiet Sunday in the harbour at Urk without a single stone being thrown. Facilities for yachts are not good, but it is an interesting town and a yacht can normally find a friendly trawler or barge to tie up to.

A short walk from Urk will bring you to Schokland, where in the old church building there is an exhibition of relics found when the polder was drained. There is also a *Hoogwaterkanon* that was fired when a very high tide was expected in order to warn the inhabitants.

The southern shore

The harbours at Lelystad and Pampushaven are werkhavens but are suitable for yachts. The island of Pampus is not a bad place for a picnic, but you have to anchor off-shore and go ashore in a tender. Due south of Pampus island lies Muiden. The jachthaven and canal are reached by entering between two moles. The eastern one is sometimes partly submerged. At night there are two leading lights flashing white, and the passage in is simple and safe provided you keep the two lights exactly in line. There is a jachthaven on the eastern side of the canal and the Royal Netherlands Yacht Club on the west side. British yachts are very welcome at the Royal Netherlands Yacht Club and the facilities there are excellent. The harbourmaster and the members of the club are most obliging.

The Randmeren

Between Zuid- and Oostelijk Flevoland and the old provinces of Utrecht and Gelderland lies a stretch of water rarely used by commercial craft which provides an excellent touring ground, more sheltered than the Ijsselmeer. The Flevoland shore is a blank dyke wall but the southern shore contains many small towns with jachthavens and moorings. From west to east these are: Huizen, one of the larger towns; Eemond, leading to the small town of Eemdijk; Nijkerk, just to the west of the first locks; Harderwijk, where yachts should use the two harbours on the southern shore and not the werkhaven to the north; and Elburg, a small square town surrounded by an equally square moat. There are also jachthavens at Roggebotsluis and at Ketelmond at the mouth of the River Ijssel.

There are a number of other mooring places for yachts on the southern shore and on the islands in the meers. These moorings, like those in the Veersemeer, are suitable for self-sufficient yachts to get away from it all. The main channel through these waters is 9ft deep and very well buoyed. For sailing craft that may have to tack, inverted triangles on posts mark the limit of the water at about the one-fathom line on the southern shore. The bridges all lift and there are speed restrictions on the water.

FRIESLAND

'A paradise for yachting enthusiasts,' is a phrase used to describe Friesland in many of the guide-books, and this is certainly an understatement. Well endowed with small lakes and criss-crossed by a network of navigable canals; bounded by the Waddenzee in the north and the Ijsselmeer in the west; a place where the ratio of boats per family is one to one; with only light commercial traffic and this largely confined to one route; with the countryside pastoral and given over to herds of the famous Friesian black and white cattle; and with small picturesque historic towns dotted freely about the province, Friesland becomes as near an ideal cruising ground for small craft as can be imagined. Most of the bridges open, so that fixed mast sailing craft can explore nearly all the province, the only limitation being that larger deep-draught sailing craft may find the waters in some of the meers and smaller canals too shallow if their draught is over 6ft.

Friesland is suitable therefore for all except the deep-draught sailing yacht and perhaps the very largest motor cruiser. Dinghy cruising is a provincial pastime and small sailing boats and motor boats will find they can go almost anywhere in the area. There are, however, speed restrictions of about 5 knots on most of the waterways and so skippers of high-speed craft must moderate their enthusiasm. The proportion of sailing cruisers to motor cruisers in Friesland is high. It is not thought odd to sail, or even tack, down a canal. Sailing boats are to be found everywhere. Sailing in a light breeze down one Friesian canal I rounded a bend and decided to fly my spinnaker. It went up perfectly and set beautifully, and to my surprise and delight brought a spontaneous cheer from a bargee going in the opposite direction. That is Friesland!

A few words, however, before you venture on to these waters. The Frisians are a strongly independent people, a sort of Dutch Welsh. They have their own language which, though taught in few schools nowadays, is spoken by many, though Dutch is spoken by all and most Frisians speak English as well. They have their own flag of blue and white diagonal stripes with red water-lily leaves on the white portions. Visiting yachtsmen would be well advised to purchase a Friesian flag and to fly it as a courtesy ensign when visiting this province. You should also take a supply of small change in 25 cent pieces, for this is *klompen country*. This is an old custom still practised in Friesland: when a bridge is opened for you, the bridge-keeper does not keep you waiting to pay your dues but lowers a *klompen* (clog) on a piece of string attached to a pole for you to place therein a token of your appreciation. It is customary to subscribe 25 cents, but as no change is given it can be a very expensive passage if you have not forearmed yourself with suitable coins. In general, bridges open during daylight hours during the summer but of recent years the bridge-keepers have won the right to have their lunch in peace. So if you are approaching a bridge about midday you might just as well pull into the side of the canal and have your lunch too. There are plenty of jachthavens in Friesland and, except on the Prinses Margrietkanaal, you can always tie up safely to the canal bank. These banks are often reed beds and unless the wind is exceptional the local practice of driving your boat head on into the reeds and just taking one bow line to the shore has a lot to recommend it. The reed beds will fender your craft from the bank. There are many firms in Friesland that offer boats for hire and the range of craft available is extensive.

One other local practice is worth mentioning and should be borne in mind by those cruising with motor yachts or sailing yachts under auxiliary power. Towards evening, dinghy and day boat cruisers who have run out of wind and have a long row to a suitable mooring site will hail passing cruisers. It is common practice to take them in tow and I normally make coffee, which seems to be much appreciated. In general these dinghy cruisers will not want to go to a jacht-

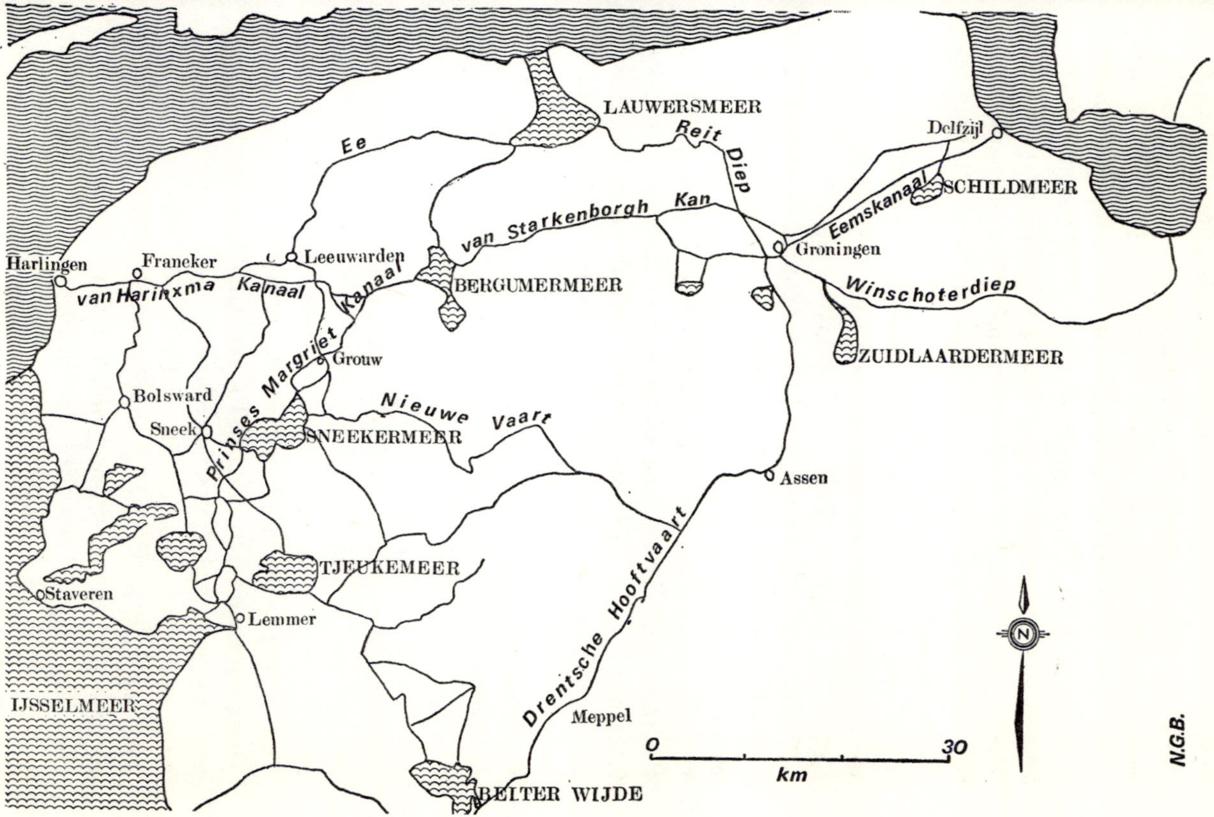

The canals and lakes of Friesland

haven but will wish to be cast off at some place along the canal.

The route used by most of the commercial traffic is the Prinses Margrietkanaal, which runs from Lemmer in the south-west on the shores of the Ijsselmeer through the Grote Brekken and the Sneekermeer, then on to the Bergumermeer before passing through the province of Groningen to Delfzijl. The depth of this canal is never less than 9ft, and channels of this depth are dredged and buoyed through the meers. Outside these channels, which are often narrow, the water may be of insufficient depth for larger barges. It is thus up to the cruiser with his shallower draught to keep out of the way of the barges, for they have little room to manoeuvre.

For the sake of convenience I have included in this section a number of meers and canals that are not actually in the province of Friesland, but in neighbouring Groningen, Drenthe and Noordwest Overijssel, where the waters afford the same type of cruising as those in Friesland. For easy reference I have taken these meers and those in Friesland alphabetically and since something has to be left out in a book of this size, I have included only canals that form direct routes between these meers or lead to some place of particular interest.

The Belterwijde and Beulakerwijde

The Belterwijde, which is divided by a dam carrying a road, and the Beulakerwijde with a number of small islands, form an area of about 6 square miles of cruising water just to the west of the town of Meppel in the province of Overijssel. These waters are only suitable for very shallow-draught boats and any boat that draws more than 3ft will probably spend more time extricating itself from the shoals than in cruising. The meers are well endowed with jachthavens and camp sites, so they are best for cruising with a dinghy or a trailed boat, although they are accessible for smaller motor cruisers and sailing yachts with shallow draught from the Ijsselmeer along the Zwartemeer and the Meppeler Diep.

Page 53 (*above*) The jachthaven at Vlissingen; (*below*) a quiet mooring on the Veersemeer

Page 54 (*above*) Holiday cruising, traditional Dutch style; (*below*) a *zeeschow* reaching across the Ijsselmeer

Bergumermeer

A little under $1\frac{1}{2}$ square miles in extent, the Bergumermeer is about 10 miles to the east of Leeuwarden, the capital of Friesland. It is also on the main barge route from Lemmer to Delfzijl. For the most part the meer has a depth of about 4ft, although there are shoals of hard sand towards the edges. The surroundings are very pleasant but there are no jachthavens. A delightful place to anchor for a meal or for a swim. South of this meer, a short canal leads to De Leyen which is shallower still, being between 2ft and 5ft deep, but it is dotted with tiny islands and is a favourite swimming place. If you are in no hurry and your boat draws too much water it is well worth anchoring in the Bergumermeer and going down to De Leyen in the tender.

Fluessen and connected lakes

The Fluessen and the smaller meers, the Morra, Oorden, Heegermeer, Grote Gaastmeer, Zandmeer, Vlakke Brekken, and Oudergaster Brekken form an interconnected series of lakes in the south-west of Friesland, about 6 miles north-east of Staveren. Depth here, apart from a few through channels, is about 6ft in the Fluessen and Heegermeer and about 4ft to 5ft elsewhere. On the north shore of the Heegermeer stands the small town of Heeg, where every other person in the summer seems to be dressed in sailing clothes. There are jachthavens at Heeg and in the south near the town of Koudum. This latter jachthaven is attached to a caravan and holiday chalet site and yachtsmen share the excellent facilities with their less aquatic fellows. If this should conjure up in your minds serried ranks of caravans in a windswept field I suggest you call in. Both my crew and I were most impressed with the landscaping and obvious thought that had gone into providing the maximum facilities and privacy with the minimum intrusion on the surroundings. So well has it been done that it took us half an hour and much walking to find the bar, which was in fact less than five minutes' walk from the boat. There are mooring places for yachts in the other smaller meers, but nothing that could be called a jachthaven. Of course, if you have the right sort of boat you can tie up almost anywhere to the bank. The surroundings are very pleasant country and the whole area is used extensively by the Dutch as a holiday and yachting centre.

Pikmeer

The Pikmeer, the Wijde of Peanster Ee and the Sijtebuurster Ee, form a trio of lakes with a total area of about half a square mile, near the town of Grouw. They are all very pleasant, having a number of small canals leading off them which are well worth exploring. The depth is rarely less than 5ft. The Prinses Margrietkanaal crosses the Pikmeer near Grouw. The town of Grouw has three jachthavens and mooring places for yachts are scattered all over these three meers. Good waters if you just want to potter about and relax.

Prinsenhof

North-east of Grouw is the Prinsenhof which covers nearly a square mile, but the water is shallow and a large section of it is reserved for rowing boats. There are however facilities for yachts, and many small islands to explore. There are swimming facilities and camping grounds and the area is well known in Holland for watersports, but there is not much room for a big boat.

Schildmeer

About 10 miles east-north-east of Groningen lies the Schildmeer, which to be honest I have never visited since it is not really on the way from anywhere to anywhere. For the sake of the record it is a bit over 1 square mile in extent; can be reached from the Eemskanaal via the Groeve; has a mean depth of about 3ft to 5ft; berthing for yachts along the banks and also a jachthaven. The surroundings are rural.

Slotermeer

The Slotermeer is an oval stretch of water about 4½ square miles in extent some 6 miles north-west of Lemmer. Apart from the two channels which are about 6ft deep the rest of the Slotermeer is only about 4ft deep and it shallows off towards the banks. It is mainly used as a through route from Sloten to Woudsend but also serves as excellent sailing water for shoal-draught boats.

Sneekermeer

The Sneekermeer is the most famous sailing meer in Holland, with an area of just over 2 square miles. For most of the holiday season, and particularly during Sneek Week, the third week in August, the meer and the neighbouring town of Sneek, which is reached through the Houkesloot in a distance of about 2 miles, has a general regatta air about it. The *Eerste Friese Zeilschool* is on the Houkesloot and there are always large numbers of young people dinghy and day-boat cruising. The meer is well provided with mooring places and camp-sites and there are two good jachthavens to the east of the town of Sneek. A main through barge channel runs across the meer, but there is plenty of water for all. The depth of the meer outside the barge channel is about 8ft to 10ft. The surroundings are pleasant, rural and wooded on the eastern shore.

Tjeukemeer

The Tjeukemeer is a large meer some 13 square miles in extent, lying about 5 miles north-north-east of Lemmer. There are two main channels across it, both with a depth of about 6ft, but extensive shoals lie to the east, so that vessels with a draught over 3ft will find these waters restricted. The meer is reached from Lemmer through the Grote Brekken and the Follega Sloot. A new road is being constructed and this has a fixed bridge with a height of about 36ft under the spans, so tall yachts are unable to pass. The meer is set in pleasant surroundings but it is not on the way to anywhere in particular and is not often visited by yachts. It is best regarded as a place to go if you have time on your hands. There are no mooring places for yachts nor jachthavens at present, but you can always drop anchor anywhere off the channels.

Zuidlaardermeer

Just over 3 square miles in extent and lying about 10 miles south-east of Groningen is the Zuidlaardermeer. The way is blocked for sailing yachts with fixed masts by a fixed railway bridge with a headroom of 10ft. The waters have shoals and the meer is not on a route to any place of interest to a holiday cruiser. A quiet backwater for those who want to be alone.

The canals of Friesland

The canals of Friesland are far too numerous to mention individually, but they are all in excellent repair and provided, as I have previously warned, you have armed yourself with suitable coinage and don't want to travel during the lunch hour, you may cruise therein as much as you like. Most of the canals and lakes in Friesland are kept at the same water level during the summer, so once in the province you will rarely have to pass through a working lock. There are restricted opening times for bridges and locks in Friesland on Sundays, but the restrictions are not very severe. If you intend to range far and fairly fast then you should obtain a copy of the opening times from the ANWB. The neighbouring province of Groningen has much greater restrictions, so far as the opening times of bridges and locks are concerned, and anyone cruising that way should travel during the week. Some of the bridges over the main Prinses Margrietkanaal do not open at all on Sunday and it is possible to be shut in between two bridges, whose closed headroom may be as low as 3ft, for a whole day. This happened to me one year when I was travelling to Delfzijl, and you have the alternatives of fuming for twenty-four hours or accepting things philosophically. Such bridges do usually open on Monday morning at about five minutes after midnight.

Some canals in Friesland deserve special mention because they have some peculiar character of their own to recommend them to the cruising yachtsman; because they link places of special interest within the area; or because they link places already mentioned in this section.

Connecting routes

The easiest way from the Slotermeer to the Heegermeer is through the Ee to Woudsend, a quaint town often with very congested waterways since it is at a crossing of two much used holiday canals, and thence to Heeg at the northern end of the Heegermeer via either the Woudsend or the Nauwe Wijmerts. All the bridges on this route are opening ones. For motor yachts that need no more than 6ft headroom there is an alternative route through wooded country by leaving at the south-east corner of the Slotermeer, passing through the town of Balk, then through the wooded area of Gaasterland and finally entering the Oude Karre at the southern extremity of the Heegermeer.

Yachts entering Friesland at Makkum have two routes to the meers. Both start along the Grote Zijlroede and the Van Panhuijskanaal, but at the Van Panhuijsbrug you can either turn south towards Workum and then east into the Heegermeer or you can turn north to the outskirts of Bolsward and then south-east to Ijlst where you can either branch north-east to Sneek and the Sneekermeer or continue south to the Heegermeer. All the bridges on these routes are lifting bridges and the commercial traffic is light. The southern route is the more direct but the less interesting.

The Sneekermeer is connected to the Tjeukemeer via the Noorder Oudweg, Langweer, where there is an excellent jachthaven and the Scharster Nieuwe Rijn. This route has little or no commercial traffic and is better endowed with scenery, but the more direct route runs through the Prinses Margrietkanaal to the Grote Brekken and then east through the Follega Sloot.

It is possible to travel from Lemmer to Urk across the Noordoost Polder instead of venturing out into the Ijsselmeer. The Lemstervaart runs south from the harbour at Lemmer to Emmeloord and is little used by commercial craft. The locks are relatively small by Dutch standards but what they lose in surface area they make up for in depth. The fall is about 30ft in one chamber, deep by local standards. At Emmeloord you turn west along the Urkervaart, another quiet canal, to the town of Urk where there is a correspondingly deep lock. You can turn east at Emmeloord and travel along quiet canals to Zwartsluis or to the Ketelmeer. An important canal often used by commercial traffic links Zwartsluis with Groningen via the towns of Meppel and Assen.

Apart from the ports on the Ijsselmeer and the town of Sneek, a number of other towns are worth visiting in this cruising area. Leeuwarden, the capital of Friesland, is on the Van Harinxmakanaal which links the Waddenzee port of Harlingen with the Prinses Margrietkanaal. Both Harlingen and Leeuwarden have facilities for yachts and are well worth visiting. Leeuwarden has many sixteenth-century buildings and is built on three *terpen*. It was a port in past times but the inlet has been silted up for many years. The Friesian Museum, the most important provincial museum in the Netherlands, is at Leeuwarden and amply repays the time spent in visiting it. Two statues in the town, one of Count Willem Lodewijk van Nassau (1584–1620) and one of a pedigree cow, are known respectively as *Us Heit* (our father) and *Us Meme* (our mother). There is an excellent municipal museum, a modern provincial library and a 'Children's Farm'. At St Anna Parochie, a few miles north-west of Leeuwarden, the pretty Frisian girl Saskia van Vylerburgh married Rembrandt van Rijn on 22 June 1634. Harlingen is still an important port, not only for the ferries to the Friesian islands, but also for coasters and a large fishing fleet. 'De Zoutsloot' here has been restored with old style bridges and gas street lamps.

Dokkum can be reached from Leeuwarden via the Dokkumer Ee and is the northernmost town in the Netherlands. It has preserved its hexagonal fortress shape and still stands complete with moats, bulwarks and bastions. From Dokkum you can reach the now enclosed Lauwerszee, renamed by the logical Dutch the Lauwersmeer, through the Zuider Ee. Farther east lies the town of Groningen in the centre of the province of the same name. The Martini tower, which can be climbed, dominates the whole town and from the top you can get an excellent view of the surroundings. There are many historic buildings in the town and the Martini church houses a monumental organ built by Rudolph Agricola.

At the extreme eastern end of the Prinses Margrietkanaal lies the important seaport of Delfzijl. Along the sea wall is a recreation area which houses an aquarium and a collection of over 4,000 different shells. There are facilities for yachts at Delfzijl but you must lock out of the canal into the harbour to reach the jachthaven. This has floating pontoons, so the fact that the moorings are tidal is of little consequence. Farther east still lies Germany, or you can sail out of the estuary between the German and Dutch Friesian islands and into the North Sea.

Throughout the provinces of Friesland and Groningen you will find many farmhouses built according to one of two traditional styles. There are 'head-neck-trunk' farmhouses, where the residential part (the head) is connected to the barn (the trunk) by a low passage (the neck), and also 'stelp' farmhouses, the large barn being directly connected to the house. Often the farmhouse is tiled and the barn thatched. It is said that the Friesian cattle breeder sets greater store by his cattle than by his wife.

It is impossible to leave Friesland without mention of Greate Pier, a peasant boy who rose to become the Admiral of the Zuyderzee. He died at Sneek in 1520 and his tomb is in the Gritekerk there. His helmet is kept in the soldiery room of the Sneek town hall.

FRIESIAN ISLANDS

Out across the Waddenzee you come to a chain of islands that have a character of their own. Pleasantly rural and with miles of sandy beaches they make excellent holiday and cruising spots. It is as well not to underestimate the Waddenzee. Tides are strong and although the waters are sheltered to some extent from the full force of the North Sea you would be well advised to venture only into this area with a well found seagoing craft. The channels are devious and the waters shallow in many places, so it is best to use a shallow-draught boat. Sandbanks constantly shift and up-to-date charts will be required, but the islands are well worth visiting. English is not so widely spoken in the islands as it is on the mainland, for until recently German was the second language taught in the island schools.

Texel

The westernmost of the islands is Texel, reached in half an hour's sailing from Den Helder. The harbour on the south-west tip of the island is for the ferry service to the mainland and yachts should tie up in the harbour at Oudeschild. This harbour is used by a large trawler fleet but there is always room for a few yachts. It is normal for yachts to moor in the Noorderhaven, although the yacht harbour is marked on Dutch charts as being in the Zuiderhaven. I have tied up in both and would suggest that you enter with care and take local advice. The Waddenzee shore slopes gently and is safe for bathing, while the North Sea shore shelves very rapidly and it is inadvisable to enter the water farther than knee-deep unless the weather is settled, the sea calm and the tide slack. The general tidal drift along the coast is strong and it is easy to be swept out to sea. The undertow from the breakers is very powerful.

On the island there are several towns of interest and it is possible to hire bicycles. This is the best way of getting about. The capital of the island is Den Burg, a fascinating town with pavement cafés and much of historical and architectural interest. Oosterend is also exceptionally interesting, while at the north of the island De Cocksdorp is a holiday resort not far from the lifeboat-station and the lighthouse at Eirland which is also worth a visit. De Koog on the North Sea coast is another holiday resort and south of this are extensive pine-woods with well-signposted bridle paths for horses and cyclists. There are plenty of small, inexpensive restaurants on the island, which is used by the Dutch as a family holiday resort.

In common with the other Friesian islands cranberry bushes abound. You should however not pick the fruit as it is reserved for the islanders who make it into excellent wine, jam and pastries. The cranberry, a native of the New World, arrived in the islands from seeds that germinated after a ship-wreck. Perhaps a thankoffering, since it was the citizens of Leeuwarden, the capital of Friesland, who were the first Dutch people to vote for the recognition of the United States by Holland.

Vlieland

Between Texel and Vlieland lies a nature reserve. The gap between the islands where the North Sea meets the Waddenzee is dangerous and vessels should on no account attempt to negotiate these waters. Since the only harbour on Vlieland is at the eastern end there is no reason to do so. The harbour at Vlieland can be reached by following the channels across the Waddenzee or, if you prefer and the weather is good, by sailing round the North Sea side of the islands and entering between Vlieland and Terschelling. I always prefer to follow the channels across the Waddenzee as I find this more

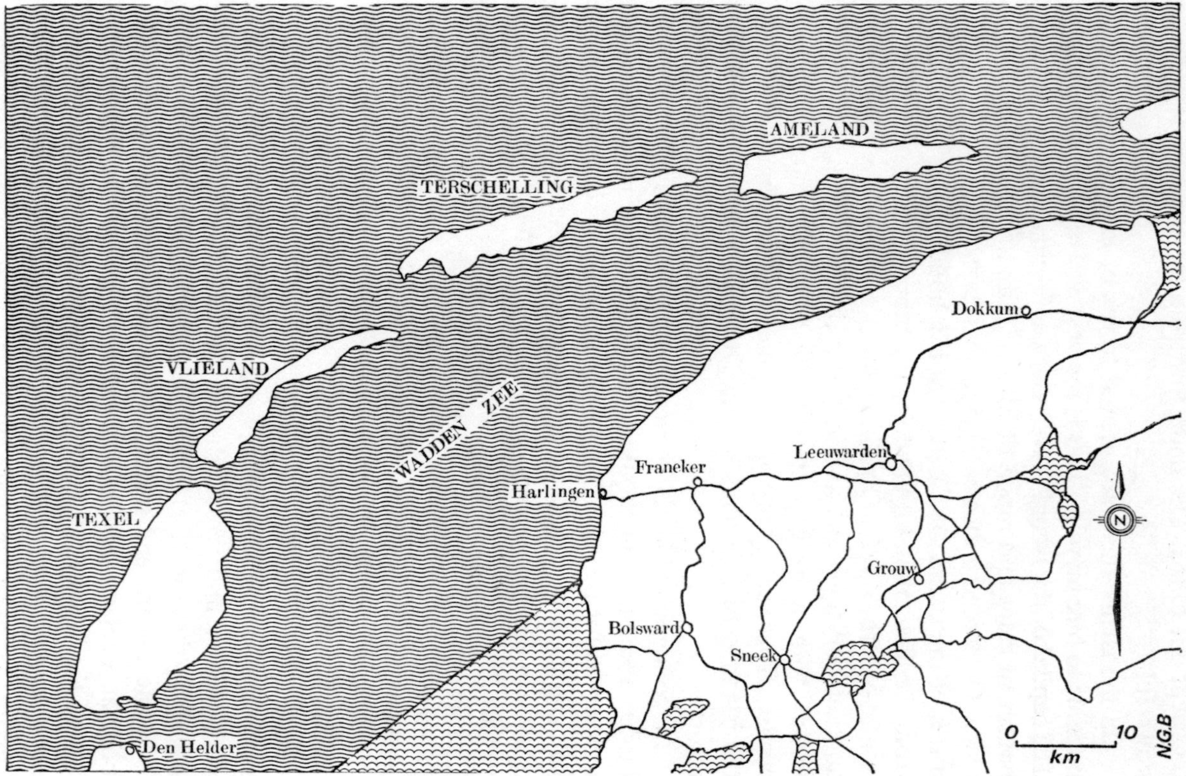

The Friesian islands

interesting. These channels are well buoyed and lead between sandbanks that dry out at low water. The tidal streams are caused by the North Sea flooding and ebbing between the islands so that to take advantage of the tide when passing from one island to another you must aim to be about mid-way at high water. This is however only a rough guide and a tidal stream chart, or a mixture of advanced mathematics, an intensive study of the sandbank formations and a fair measure of inspired guesswork is needed to be at precisely the right place at the right time. Erskine Childers in *The Riddle of the Sands*, although writing about the German Friesian islands, gives an excellent account of this type of water and the techniques required to navigate on it. Vlieland is devoid of cars and sandy beaches stretch for miles. The Waddenzee side of the island is excellent for bathing and when the tide is out a vast expanse of sand stretches out towards the main-land. On all these islands you can have half a mile or so of beach all to yourself. Vlieland is 18 miles from Harlingen on the mainland and a ferry runs a regular service. The ferry boat has the curious feature that although it has a traditional funnel, it is false, for small puffs of diesel exhaust issue from the after mast. Vlieland is 31 miles from Oudeschild on Texel.

Terschelling

West Terschelling is the harbour on the island of Terschelling, the next island in the chain. It is somewhat larger than Vlie-land and has better facilities for yachts. The town is a good shopping centre and has some historical and architectural interest. The square-built lighthouse can be climbed in the summer and gives an excellent seagull's-eye view of the town and the surrounding sea. There is a large nature reserve on the island which you may tour with a pass from the Forestry Commission or you can be taken on a conducted tour in a horse-drawn cart. Terschelling, like Texel, is much used by the Dutch as a holiday island. Relatively simple and in-expensive hotels can be found and it is also possible to book a holiday chalet or stay at one of a number of farmhouses that take guests in the summer. The Waddenzee side of the island has very good bathing beaches with sand as far as the eye can see.

Ameland and the other islands

Ameland, the next, is a holiday island with excellent holiday beaches to the south. The channels to it across the Waddenzee are well buoyed but extremely tortuous and unless you have experience of sandbank hopping, a shoal-draught boat that will take the ground comfortably if you misjudge it, time to spare and settled weather, you would be better advised to sail round the North Sea coast of Terschelling and then enter between Terschelling and Ameland. There are two harbours for yachts on Ameland and both dry out, so to visit them you need a yacht that will take the ground safely. It would be unwise to proceed eastwards to the next island of Schiermonnikoog across the Waddenzee without local knowledge and experience. There is a small harbour and a place for yachts to tie up but this dries out. The island has nothing to offer the holiday cruiser that the other islands haven't. So, unless hopping over sandbanks is your especial watersports thing, you would be best advised to return to the mainland at this point, either by sailing round to the estuary of the Eems and entering the mainland at Delfzijl, or locking in at Lauwersoog to the Lauwersmeer, or by returning to Friesland at Harlingen.

WESTERN HOLLAND

I have included in this area the whole of the province of Noordholland, Zuidholland north of Rotterdam and the extreme western part of the province of Utrecht. It is bounded on the west by the North Sea, on the north by the Waddenzee, to the east by the Ijsselmeer and to the south by the River Lek. The whole area presents a uniform type of cruising water, although divided from east to west by the Noordzeekanaal. A number of lakes in the area are suitable for small sailing craft and dinghies, but for the most part the motor cruiser is best for cruising these waters, although I have cruised most of the area in a sailing yacht, using it as a motor cruiser after having taken the mast down. That part of the area south of the Noordzeekanaal is by far the most densely populated part of Holland; therefore there are more fixed bridges and the times of opening of the movable ones are more restricted than elsewhere. With the exception of one or two canals this area is therefore not suitable for seagoing sailing yachts with fixed masts.

The canals carry a fair amount of commercial traffic, particularly in the south, but for the most part commercial craft are relatively small and barges seldom more than 1,000 tons. In this area, again largely in the south, lie many of the historic towns of the Netherlands whose names are familiar to people outside Holland, and I would venture to suggest that this is also the part of their own country that the Dutch know best. Gouda, famous for cheese; Delft, noted for its china; 's-Gravenhage (The Hague); Haarlem; Hilversum; Utrecht; Rotterdam and Amsterdam are all in this area. Here, too, lie the great bulb fields as well as a large part of the major industry of the country, so there are many good and varied reasons for visiting western Holland.

Jachthavens are by no means as numerous as elsewhere in Holland, but they are frequent enough to satisfy most cruising needs, and it is possible to find somewhere safe to tie up in most towns. A motor cruiser without too much top hamper can cruise these waters without excessive delays due to bridges but, as there are speed restrictions on most canals, there is little advantage to be gained from having a high speed craft with a possible premium on space. A reliable engine or engines are, however, a must, since the disturbance caused by the barges may require that extra burst of power and it is comforting to know that you have it.

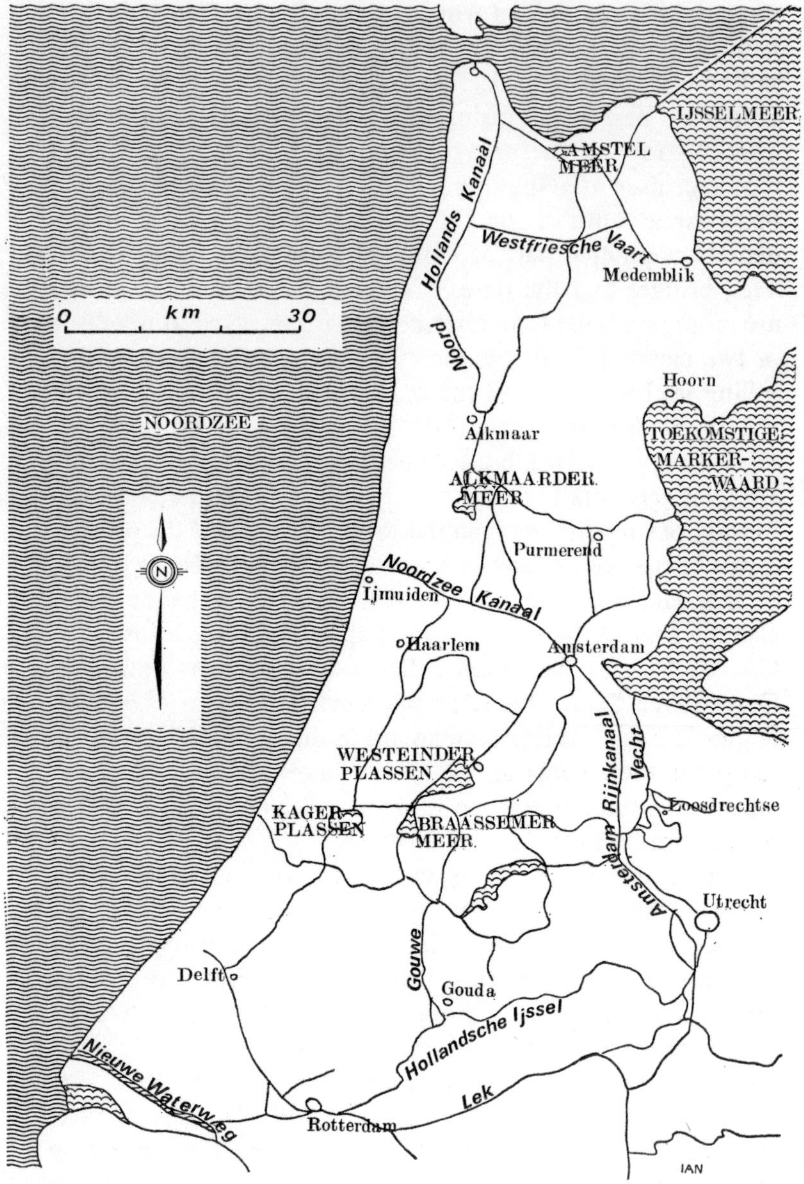

Waterways of western Holland

The canals to the north of the Noordzeekanaal are similar to those south of it, but there are fewer historical towns, a lesser density of industry and not so much commercial traffic. Even so, the province of Noordholland has a character of its own and is well worth a visit. It affords some excellent cruising waters, although jachthavens are scarcer than in the south. Sailing yachts are more frequent in this northern half of the area and the number and frequency of opening bridges is greater than in Zuidholland.

The lakes in western Holland that are suitable for dinghy cruising and for small sailing yachts are set out alphabetically below. In general these lakes are shallow and some have entrances that are either under low bridges or through shallow locks. On most there are some restrictions on power vessels and often a permit has to be obtained locally. On some, speedboats can be used provided a permit has been obtained in advance.

Alkmaardermeer

The Alkmaardermeer, nearly 1,500 acres, is reached from the Noordzeekanaal through either the Zijkanaal G or the Noordhollandskanaal. Depth is between 6ft and 8ft but there is a buoyed channel dredged to 9ft through the centre of the meer from north to south. The eastern bank is stony while the western shore is composed of reed beds in which it is prohibited to berth. The lake is used for all forms of watersports. Jachthavens exist at De Hoorne and Akersloot in the north and at Uitgeest in the south.

Amstelmeer

This is a meer of nearly 2,000 acres in extent and lies right in the north of the area. It can be reached through the Slootvaart and Westfrieschevaart from Medemblik or the Balgzandkanaal from Den Helder. The depth is in general about 6ft, but there are some shallower patches. The lake is used for most watersports and there is a jachthaven in the extreme north-east at De Houkes. Three buoyed channels cross the meer.

Braassemermeer

On the route from Alphen an den Rijn to Amsterdam, the Braassemermeer is a little over 1,300 acres in extent. Apart from the south-east and south-west corners the depth is over 6ft everywhere. It is used for all types of watersports and there are three jachthavens on the western shore and one on the eastern shore. The crescent-shaped island in the south is a favourite anchorage. The main barge channel runs close to the western shore.

The Kagerplassen

A little under 1,000 acres in extent, the Kagerplassen—four interconnected lakes—can be reached along the Zijl from the university town of Leiden. Depths vary from 5ft upwards and the whole area is given over to watersports and outdoor recreation. There are numerous jachthavens, campsites and simple inexpensive restaurants. The lakes are less than 1 mile from Sassenheim, a centre of the bulb-growing industry. There is little commercial traffic.

The Westeinderplassen

The 22,000 acres of the Westeinderplassen lie about miles north of the Braassemermeer, and are reached along the Ringvaart van de Haarlemmermeerpolder. Depth in general is over 6ft. It is an area used extensively for water recreation, chiefly sailing and small motor cruising. There are many jachthavens along the north-east bank but the jachthaven Vafamil at Vrouwentroost is a military one and should not be entered. Speedboats must not use the lakes unless a permit has been obtained. The speed limit among the islands is about 3 knots. There are no through barge routes.

Page 71 (*above*) The lighthouse at Marken; (*below*) new locks in the dyke surrounding Markerward

Page 72 (*above*) The harbour at Volendam; (*below*) the *Drommedaris* and entrance to the old harbour at Enkhuizen

Rotterdam to Amsterdam

Starting from Rotterdam there are two main routes northwards. The first is north-east along the Hollandsche Ijssel, past the storm gates at Krimpen aan de Ijssel to Gouda where you leave the tidal water for the Verbindingskanaal and the Gouwe. Gouda is a town of great charm and much history. The cheese market is genuine rather than a tourist attraction. Gouda's town hall, in the middle of the square, is a magnificent example of late-Gothic extravagance. The platform at the rear was used as a scaffold in less enlighted times. The St Janskerk with a carillon of forty-nine bells contains some beautiful stained glass windows rivalling those at Chartres. There is also a pipe museum which will be of interest to smokers. Gouda is not only famous for its special *Boerenkaas* cheese but also for its candles, ceramics (much 'Delftware' is made in Gouda), pipes, and treacle wafers. The cheese market runs concurrently with the ancient trades market, where you can see clogs, rope, pipes and treacle wafers made every Thursday during the summer months.

From Gouda the canal leads northwards through the towns of Waddinxveen and Boskoop. This part of Holland is famed for flowering shrubs and topiary. At Boskoop, not very impressive from the canal, is an excellent 'Rosarium' and botanical gardens. The canal reaches the Oude Rijn at Alphen aan den Rijn. The 'Avifauna' here is probably the best aviary in the world with some 10,000 birds of over 400 different species.

A traveller crossing over the Oude Rijn and entering the Aarkanaal will travel through pleasant countryside and no large towns until he reaches Uithoorn, where the canal joins the Amstel River. He can then follow the charming Amstel River down into Amsterdam. This is by far the most scenic route to Amsterdam. If, however, he turns west at Alphen aan den Rijn he will be able to travel to Amsterdam across the Braassemermeer and the Westeinderplassen, already described. From there northwards the canal passes between the Amsterdam Bos, a very pleasant park, and Schiphol airport before going through an extensive and not very inspired part of the city to the centre of Amsterdam.

The second important route northwards from Rotterdam leads first south-west down the Nieuwe Maas from the Royal Maas yacht club and then north into Parkhaven by the Euromast. Here a cruiser must lock in with the barges and after negotiating an industrial area will eventually come out into a short stretch of open country before reaching Delft. Do not be misled by the pretty rural scenes on the delftware. Delft is an important industrial town and from the canal has nothing to recommend it. There is no jachthaven and I have never found anywhere suitable to stop. However, the town is steeped in history and delightful to the eye when you get away from the through canal, so it is worth visiting even if you have to leave your boat elsewhere and travel by bus.

Founded in 1653 the Royal Factory at Delft, employing about 300 artists, is the only one remaining of the 30 potteries that existed in the town in the seventeenth century. It is possible to visit it and see genuine Delftware being made. This has a triple signature of a vase topped by a straight line, the letter 'F' below it and the word Delft. Prices vary according to the complexity of the design rather than the size of the article. There are many other designs besides the well-known blue and white. The Antique Dealers Fair at Delft attracts many visitors. More modern is the Technological University. The new Auditorium building is a marvel of concrete construction. The Oude Kerk has a slanting steeple, and houses the tombs of the Dutch heroes Piet Hein and Tromp. There is also a rather modest monument to Antonie van Leeuwenhoek, the inventor of the microscope and discoverer of microscopic organisms. Delft is often called 'The loveliest little city in the Netherlands', but you must leave the canal to see its real beauty.

North of Delft the canal leads close to 's-Gravenhage, the

governmental, though not the royal, residence. Unless you like gazing at buildings which house government departments there is not much of especial interest to those on holiday. The VVV can provide tourist information. The residents are rather different from the inhabitants of the rest of the Netherlands and many even of their fellow countrymen do not understand them. Despite all this the Mauritshuis is one of the greatest art galleries for its size in the world and houses a superb collection of Dutch seventeenth-century paintings, including fifteen by Rembrandt, and some of the work of Vermeer and Frans Hals. 's-Gravenhage is worth visiting for this alone.

The canal leads northwards through rural surroundings to the very beautiful city of Leiden, the birthplace of Rembrandt. The university, dating from 1575, is the oldest in the Netherlands. There are specialised museums in the city devoted to ethnology, natural science, the army and weapons, also botanical gardens and an observatory. The city is connected historically with the Pilgrim Fathers and a plaque commemorates the place where William Brewster had his Pilgrim Press. It is possible to penetrate the city by canal and you can safely leave your craft for short sight-seeing trips. For a longer visit you would be better advised to go to the jachthaven on the main canal just north of the city or to tie up at one of those in the south of the Kagerplassen, preferably at Warmond, which has better communications with Leiden.

Northwards again the canal leads to Haarlem, more residential and industrial than historic. There is a jachthaven in the sportspark in the south of Heemstede, a suburb of Haarlem, and others north of Haarlem at Spaarndam and Penningsveer. The canal then joins the Noordzeekanaal about halfway between Amsterdam and the sea. There is a fair amount of commercial traffic on this route from Rotterdam to Amsterdam and it is far less scenic than the route through Gouda, but it does pass a number of interesting towns and cities. Sailing yachts with fixed masts must take the route through Gouda as there are now fixed bridges on the route through Leiden.

Rotterdam to the Ijsselmeer

From Rotterdam a very interesting route runs up the tidal Lek, though you will need a reasonably seaworthy craft as there is a fair amount of barge traffic, but the scenery is good and for the first part of the route there are no bridges, lifting or otherwise. There are, however, a number of ferries and these should be given right of way. Several good jachthavens are to be found, notably at Schoonhoven and Vreeswijk. Schoonhoven is famed for its small silverware workshops, many of which welcome visitors. At Vreeswijk you can continue up the Lek, a route which is dealt with later, or turn north into the Merwedekanaal. It is not advisable for yachts to use the Lekkanaal. The Merwedekanaal brings you out into the Amsterdam–Rijnkanaal; except on Sundays this is not suitable for yachts, as it is a sort of motorway for enormous barges.

If your craft requires no more than 9ft headroom and has a draugh tof less than 6ft you can cross the main canal and cruise through the historic town of Utrecht, and then follow the quiet and very beautiful River Vecht which winds its way northwards to enter the Ijsselmeer at Muiden. This is a delightful rural cruise apart from the urban area of Utrecht. Utrecht can be avoided, if you have a sound boat, sounder nerves and an engine that is not going to falter, by taking a short trip down the Amsterdam–Rijnkanaal to the first sluice north of Utrecht and entering the Vecht at this point. You will have travelled only about 7 miles on the Amsterdam–Rijnkanaal. Personally I would rather go through Utrecht which, although from the Amsterdam–Rijnkanaal it looks like a dormitory town of uninspired modern blocks of flats, turns out to be a charming and very interesting city. Apart from the usual galleries and museums and historic buildings,

Utrecht has a 'Dutch Gold, Silver and Bells Museum', and the fascinating National Museum 'Van Speeldoos tot Pierement' (from musical box to barrel organ). The Dutch Railway Museum is also at Utrecht, a must for railway enthusiasts of all ages.

The Noordzeekanaal

The Noordzeekanaal runs from the North Sea port of Ijmuiden to the Ijsselmeer at a point just east of Amsterdam. It is used as far as Amsterdam by ocean-going ships and so a passage through it should not be attempted in anything less than a good motor boat or a seaworthy yacht. There are no fixed bridges, all being lifting ones which have a height of at least 35ft without the span being raised. The banks of the canal are less industrial than those of the Nieuwe Waterweg to Rotterdam, since the only industrial areas on the Noordzeekanaal are at the two ends around Zaandam and Ijmuiden. The locks at Ijmuiden lead into the outer tidal harbour and then into the North Sea. Any craft wishing to shelter in the outer harbour before putting to sea should choose one of the two fishing harbours, preferably the Haringhaven, but it should be remembered that this is not a jachthaven. Yachts are therefore not officially welcome and the fishing fleet that uses the harbour is both large and busy. A yacht or motor boat entering from the sea should go to the smallest and southernmost lock, the Zuidersluis, some 350ft long and 50ft wide. The lock system at Ijmuiden is the largest in the world and the biggest lock, the Noordersluis, measures 165ft × 50ft × 1,320ft. A new jachthaven exists just to the east of the Noordersluis and a yacht wishing to stay at Ijmuiden should berth there. Ijmuiden is, however, an important industrial town and a busy port and fishing harbour, so a protracted stay is not recommended. There are no jachthavens between Ijmuiden and Amsterdam so that a yacht not making the whole length of the canal in one sailing should pull off the main canal at Zijkanaal C and stay at the jachthaven at

Spaarndam, or should enter the Zijkanaal D and stay at the jachthaven Nauera. I have stayed at the jachthaven marked on the ANWB charts at Zaandam, but it turned out to be a rowing club. I was however made most welcome. They are however unable to cope with larger yachts.

Amsterdam

There are two jachthavens at Amsterdam that are suitable for visiting craft. The first is Sixhaven, directly opposite the Centraal Station and just east of the entrance to the Noord-hollandskanaal. Although from the canal it looks as though you should enter this jachthaven from the western side this is not so. The only entrance is from the east. If you are coming up the Noordzeekanaal from the sea you have to pass the entrance to the Noordhollandskanaal and the ferry before crossing the canal and entering the harbour. There are no facilities at Sixhaven, except a place to tie up and water, but it is very centrally situated and a free ferry runs frequently day and night to the southern bank where is the main tourist area of the city. The other suitable jachthaven is in the eastern section of the Joh van Hasseltkanaal. Yachts of up to 30ft with a draught of less than 6ft can shelter here at the jachthaven Z V Aeolus. There are both fresh water and toilet facilities, and it is only a short bus ride to the centre of the city. Visiting craft will notice another jachthaven on the south bank, but this is not recommended since most of the berths are not available for visitors and although some protection is gained from the piles, yachts moored there are constantly rocked by the wash from passing ships. Amsterdam is an exceptional city and a tourist paradise; the VVV near the station will give you all the help and advice you need. If you haven't seen enough water by this time a trip on one of the sightseeing water buses is well worth the modest charge.

Noordhollandskanaal

Cruisers wishing to go north of Amsterdam can either lock

out into the Ijsselmeer at the Oranjesluizen or travel northwards through the Noordhollandskanaal. This canal passes through the northern residential part of Amsterdam and is used by a number of commercial craft as well as yachts. The density of commercial traffic is, however, light and the cruiser soon finds open country, a good canal and few bridges. The canal leads between Wormer and Purmer, two large areas of reclaimed land, to the town of Purmerend where there is some industry and a set of locks. Immediately north of these locks the canal takes a right angle turn and yacht skippers should keep a sharp look-out for craft going in the opposite direction. The canal then passes through more reclaimed land and on to the northern edge of the Alkmaardermeer, before turning north again to pass close to the interesting northern town of Alkmaar. The cheese market held here every Friday morning in the summer is world-famous and a great tourist attraction. The cheese carriers, members of an ancient Guild, carry the cheeses, piled high on large wooden stretchers, from the warehouses and barges. The weigh-house has a clock where at each hour two small horsemen appear and ride round while a horn blower plays a tune. The town also has several historic buildings, a museum and a deer park.

Northwards of Alkmaar the canal passes very close to the North Sea, being separated from it by a range of sand-dunes and a few small towns. Towards the extreme north of the canal there are a few floating bridges, which the bridge-keepers winch in towards the bank to let boats pass. They live in houses close to the bridges, and some form of horn is desirable to let them know that you are waiting, since water-borne traffic is light and the bridge-keepers are often engaged in digging their gardens. At the northern end of the canal there is a jachthaven on the canal side of the locks. This seems to be run exclusively for club members, and I have to admit that although I was found a berth and not charged for my stay I somehow expected a warmer welcome. The town of Den Helder has little to offer the tourist, but for the yachtsman it is an important jumping-off point for a visit to the Friesian islands. Yachts locking out at Den Helder should use the smaller eastern locks. In the tidal harbour there is another jachthaven run by and for the Royal Netherlands Navy. Visiting British yachts are made most welcome here, but accommodation is limited. It has to be emphasised that Den Helder is a naval town and this jachthaven is in the naval base, so visiting yachtsmen must keep to the rules and most decidedly must not take photographs.

Other canals in Noordholland

The other canals in Noordholland are small and have little commercial traffic. Medemblik on the Ijsselmeer is connected to the Noordhollandskanaal at a town called Stolpen. The canal passes the town of Schagen where the West Friesian market, with folk dancing, is held every Thursday in the summer months. The waterway is quiet and the surroundings pleasantly rural. Edam is also connected to the Noordhollandskanaal at Purmerend and after passing under several bridges in Purmerend, all lifting ones, the canal is unobstructed until you reach Edam, where there are seven lifting bridges in a distance of less than a mile. Monnickendam can also be reached from the Noordhollandskanaal by using the Trekvaart van het Schouw naar Monnickendam. This route is pleasant and rural but has one fixed bridge with headroom of about 12ft. Half-way along the Trekvaart, at Broek in Waterland, a branch takes you to a series of small lakes about 3ft deep that lead to the villages of Zuiderwoude, Uitdam and Holysloot. These lakes are all used for cruising and if you have a shallow-draught boat they are worth exploring.

The Alkmaardermeer can also be reached from the Noordzeekanaal by the Zaan through Zaandam. This route is shorter than the route using the Noordhollandskanaal but for over half the length the banks, particularly the western, are industrialised, so unless you are in a hurry there is no point in using this route. The Zijkanaal D forms the final route from the Noordzeekanaal to the Alkmaardermeer. The area is not industrialised and the bridges, all lifting, are few. The depth of water in this small canal is about 6ft and the trip is interesting, in that it passes through an area still being drained. It is also possible to leave the Noordhollandskanaal at Purmerend and travel round the Beemster Ringvaart, about 5ft deep, and so circumnavigate the Beemster which is a recently reclaimed tract of land.

A number of smaller canals, particularly in the north, are all navigable. Most of them are quiet but have enough depth for the average cruising boat. All canals in this area have a speed limit, dependent on the particular canal and the size of craft. A good general rule is to keep at 5 knots or below. In summer bridges and locks work during daylight hours with some restrictions on Sundays, but since this is a holiday cruising area particularly for people from Amsterdam, the Sunday restrictions are not very severe.

THE GREAT RIVERS

Before you start thinking of a cruise on Dutch rivers in the same terms as you would think of a quiet cruise on the Thames above Teddington let me put the record straight. The Ijssel, the Maas and the Waal are *great* rivers. They connect sea-ports including Rotterdam, Europe's largest port, with the industrial area of Germany. They are all busy arteries of commerce, and barges of 2,000 tons and upwards are frequent users. Before you gloss over the idea of a 2,000-ton barge as being large, let me put it into perspective. Such a barge is four hundred times the size of the yacht in which I did the research for this book. Such barges can be as long as 100yd, that is about the length of a rugby pitch; as wide as 30ft, greater than the frontage of most houses in Britain. They can have a draught of 9ft, greater than the height of a single-decker bus. Couple with these dimensions the fact that these barges often battle with a 12 knot current on the Rijn and win, and you will get some idea of the enormity of these monsters of the waterways. It is essential to have a healthy respect for them and to give them a wide berth at all times.

Although some Dutch yachtsmen sail these rivers and even tack against the wind and tide, they have well-found shallow-draught yachts that can snuggle in between the groynes that protect the banks and allow the barges to pass. They have special mast-lowering gear for passing under bridges quickly and safely and above all, powerful auxiliary motors to get them out of trouble. They are also the local lads who know the waters backwards. Visiting yachtsmen should not attempt to emulate them. Sailing yacht skippers would be well advised, if they wish to cruise these waters, to leave their masts behind, use their auxiliaries and behave as though they were driving motor cruisers. If however they have only small 'get-you-into-harbour' engines they would be better off in some other cruising area. This part of the country is the province of the powerful motor cruiser and even these may have problems unless they are seaworthy and handled with a great deal of understanding and skill. To cruise these waters I forsook the comforts of my own sailing yacht, even though it had been modified to cope with the conditions on most Dutch waterways, for a faster and more manoeuvrable power boat.

Two other points should be mentioned before these waters are considered in detail. Firstly, special regulations govern the conduct of traffic; these are given in a later chapter and it is important to understand them. Of especial interest to those cruising is the regulation that a barge carrying a blue flag going upstream has right of way over any craft going downstream, and that any barge carrying a blue flag to starboard requires a ship meeting it to pass it to starboard. There is also a rule requiring that apart from these situations motor vessels must be overtaken on their port side. All ships under 15 tons, and that includes most cruisers, must give way to larger vessels. Secondly, there are a few unusual ferries.

The great rivers

Most people who have cruised on inland waters have met ferry boats, but on these rivers *kabelponts* (cable ferries) and *gierponts* (pendulum ferries) exist. The *kabelpont* propels itself from bank to bank by winding in a cable and paying it out at the other end. On some, the cable is taut across the river when the ferry moves, in others the cable is picked up from the bed, so it is safe to pass astern. It is wise to proceed with great caution until you have found out which type it is and if possible let some other boat go ahead just in case your calculations are incorrect. The pendulum ferry works as follows. The ferry is moored some way upstream from the landing stages and is joined to the mooring buoy by a cable or chain. This cable is supported on a number (usually three) buoys or boats. The ferry then swings from side to side across the river under its own power, with the mooring acting as the pivot and the ferry as the bob of the pendulum. You should therefore pass such ferries on the side of the river away from the landing stage to which the ferry is moored and not recross the river until you are upstream of the mooring buoy. If the ferry is in motion you should wait. Otherwise, a good dose of common sense, discretion and basic seamanship together with a well-found motor cruiser (the bigger and more powerful the better), and the great rivers are a delight for you to explore.

The River Ijssel

Starting from the mouth of the River Ijssel where it empties at Kampen into the Ijsselmeer in the north and proceeding upstream to Ijsselkop you cruise through largely rural scenery on a broad river with few towns of any size and a modest current in the summer of just over half a knot at the mouth to about 3 knots at Ijsselkop. The river is about 75 miles long; the width varies from about 540ft at Kampen to about 250ft at Ijsselkop. Because of a few shallow patches larger barges are unable to use the Ijssel. Kampen stands at the mouth of the river and is both a good shopping centre and a tourist town. The Dutch, noted for their chiming clocks, have carried their art here to the extreme of having a clock that chimes every seven and a half minutes. The late-Gothic town hall has a leaning tower, spiral chimneys and a pillory that is still intact. Many of the old town gateways have been beautifully restored and there is much of architectural interest dating from the town's heyday in the fourteenth and fifteenth centuries. Fuel and berths for yachts are available. The town has the first of the ten bridges to be found on the river, with about 29ft headroom under the navigational span. Some bridges higher upstream, however, are movable, with far less lowered headroom.

The lateral system of buoyage is used on the river except that middle-ground buoys are all red/black striped. The channel is also marked with *bolbakens* and the ends of the groynes with *kribbakens* (see Chapter Five). The river bypasses the town of Zwolle, which dates back to 1040, but this can be entered by locking into the Willemsvaart just after the first of the three bridges that lead to the town. About a mile and a half further upstream on the opposite bank is the attractive old fortress town of Hattem. Cruisers wishing to visit Hattem should pull off the main canal and enter the Apeldoornskanaal which, by the way, no longer leads to Apeldoorn from this end.

Between Hattem and the large town of Deventer there are a few isolated factories, several jachthavens and pleasant slightly hilly banks to the river. There is a small jachthaven at Deventer and although the town, from the river, looks like a combination of large blocks of flats and industrial buildings, it improves as you leave the water and explore. Deventer is in fact a very old town, the crypt of the church dating back to 1040. There are many old buildings and the Athenaeum library has a fine collection of old manuscripts. The Munttoren complex is a cultural centre with many exhibitions, mainly of modern art. The museum 'de Drie Haringen' has some delightful antique toys and costumes.

Higher upstream is Zutphen, its industry concentrated on the down-river side of the town. The jachthaven is to the south in the non-industrial part. Zutphen is an ancient town with some old gateways still intact. The medieval library of chained books in the St Wallburg Church attracts many visitors. It was at Zutphen that the gallant and romantic Englishman Sir Philip Sidney died in 1586 helping the Dutch against the Spaniards. Further upstream is Doesburg, situated on a bend in the river. Doesburg is an old fortified trading town with most of its fortifications still intact. Cruisers visiting the town should lock into the Gekanaaliseerde Oude Ijssel and tie up at the jachthaven there. The Ijssel finally reaches Ijsselkop just south-east of the branch with the Neder Rijn at famous Arnhem.

Arnhem is an interesting town with excellent shopping facilities and a jachthaven off the Neder Rijn. Most Englishmen remember Arnhem for the courageous but unsuccessful airborne operation in September 1944. The eight-day battle in which 7,000 airborne troops died is not forgotten. The devastated Rijn bridge that they hoped to capture remains unrepaired. The Airborne cemetery, monument and museum are places of pilgrimage. Much of the city was destroyed in the battle, but the Dutch have restored those parts of historic or architectural interest and at the same time have created a traffic-free modern shopping centre that would be the pride of any town, even one several times Arnhem's size. On the outskirts of the town are a number of parks, including the 75 acre Netherlands open air museum, the third largest zoo in the country and a 60 acre 'safari' park with lions and other jungle animals. The area to the north of Arnhem is hilly and well wooded, and at Rozendaal is the headquarters of the International Institute of Castles with models and photographs of castles from all over the world.

The Neder Rijn

The Neder Rijn runs down from Arnhem to the junction of

the Amsterdam–Rijnkanaal, after which it is named the Lek and flows on through a lock just north-east of Vreeswijk, and then to Rotterdam. I have dealt with the part from Vreeswijk to Rotterdam and so will only mention the upper reaches here. Bridges are few and all are fixed. The least headroom is about 30ft, so some sailing boats with fixed masts would be able to cope provided they sailed only with a following wind, since due to the commercial traffic, to tack in this stretch of the river is not a reasonable proposition. The current in the upper reaches is modest, being rarely above 2 knots in the summer, and in the lower reaches the tidal flow is seldom above 3 knots. The river is about 425ft wide at Arnhem and over 750ft below Vreeswijk, although the navigable channel is only about 150ft on average. The beaconage and buoyage is according to the the lateral system. The Neder Rijn is about 33 miles long, the Lek a little longer.

Apart from the jachthaven at Arnhem in the entrance to the Nieuwehaven there are also mooring places for yachts on the northern bank of the Neder Rijn, just downstream of the town. From Arnhem, going towards the sea, the scenery is rural and there are many wooded hillocks to the north including land rising to 300ft, which is quite a hill by Dutch standards. The river winds somewhat, and there are numerous groynes, all marked with beacons or buoys, to direct the course of the river and ensure that the navigable channel is as self-dredging as possible. There are several motor *gierponts* (pendulum ferries) as well as a larger number of conventional ferries. It is forbidden to berth along the river between the following kilometre posts: 878 and 879, 882 and 884, both in Arnhem; 885 and 886 in Oosterbeek, 972 and 930 at the crossing with the Amsterdam–Rijnkanaal; and between 987 and 989 at Krimpen a/d Lek.

Crossing with the Amsterdam–Rijnkanaal

The river crossing with the Amsterdam–Rijnkanaal is very dangerous and warning lights have been installed on both the canal and the river. Since, however, the canal has a lock each side of the crossing, canal craft have not usually much speed when they cross the river and there are periods when none are crossing. Approaching the crossing from either the Neder Rijn or the Lek, all signals and warning boards are to starboard. First there is a channel depth warning and then a warning about the canal crossing, then the canal crossing lights. The light signals are as follows:

One flashing yellow light	Vessels from the Prinses Irenesluis approaching
Two flashing yellow lights	Vessels from the Prinses Marijkeslius approaching
Three flashing yellow lights	Vessels from both locks approaching

The whole distance from Rotterdam to Arnhem is about 75 miles. There are a number of jachthavens on the way and the rivers form the route of a delightful cruise from Rotterdam through Arnhem and down the Ijssel to the Ijsselmeer.

The Waal

There is an alternative route from Arnhem towards the sea, though it passes through less interesting countryside. This is via the short Pannerdenskanaal which leads east of Arnhem and joins the Neder Rijn with the Boven Rijn and Germany. The Pannerdenskanaal has no fixed bridges, and if you turn west at the end of the canal you will be in the Waal which flows past the towns of Nijmegen, Tiel and Gorinchem, where the river changes its name to the Boven Merwede, then divides into the Beneden Merwede which leads to Dordrecht and the Nieuwe Merwede, leading to the Hollandsch Diep at Willemsdorp. There are few fixed bridges and these have headroom of 45ft, except at Dordrecht and Willemsdorp where the bridges are much lower. Several good jachthavens are to be found on these rivers and the surroundings are largely rural. The Amsterdam–Rijnkanaal leads into the

Page 89 (*above*) A typical Friesian farmhouse; (*below*) a modern cantilever bridge

Page 90 (*above*) Yachts tucked in amongst the trawlers at Oudeschild; (*below*) the author's yacht baulked by a bridge in Groningen. No Sunday opening

Beneden Merwede at Tiel. The rivers are buoyed according to the lateral system. The Waal is about 53 miles long, the Boven Merwede about 5 miles and the Beneden Merwede 10 miles. The rivers are all tidal and the maximum flow is about three knots, with the ebb being stronger than the flood. Width is about 900ft and the least depth about 9ft. They are all used by large commercial craft but are not overcrowded.

There is unfortunately no jachthaven at Nijmegen, but the town, which dates back to Roman times, is worth a visit. The emperor Charlemagne lived here at Valhof. Part of the town was destroyed in 1944 and, like Arnhem, some has been rebuilt and some redesigned in the post-war style. Somehow the Dutch manage to rebuild parts of their towns without destroying their soul. The country surrounding Nijmegen is very beautiful with hills and forests. Two other towns on the river are Tiel, industrialised but famed for its glass, and Gorinchem, once an old fortified town and still with parts of its ramparts intact.

The Maas

West of Nijmegen is the Maas–Waalkanaal which leads into the River Maas from the Waal. The Maas is a smaller river than the Waal and flows westward from this point to Willemsdorp, canalised at first but tidal later on. Bridges are a little lower than on the Waal and there are two locks. Jachthavens exist and for the most part the scenery is hilly and rural. This stretch of the Maas flows past the towns of Oss and 's-Hertogenbosch, where stands St John's Cathedral, often called the Netherlands' most magnificent church. There are several short canals which connect the Maas below the canalised section with the Waal, seawards of Lith. These connecting waters are much used by yachts, the two main ones being near the towns of Kerkdriel and Aalburg. The former is much used as a watersports area while the latter is slightly less important. The Maas is finally named the Amer before joining the Hollandsch Diep.

The Biesbosch

Between the Nieuwe Merwede and the Amer, just to the north-east of Willemsdorp, lies an area of creeks and woods so much used for watersports that a special detailed chart is published by the ANWB, the Biesbosch. The waters in general are fairly narrow and shallow, about 2ft to 3ft deep at low water. This is, however, the largest area of water leading from the great rivers that is suitable for small boats and cruising dinghies. At high tide larger pleasure craft can enter the area. There is no commercial traffic.

Limburgse Maas

If instead of turning seawards on entering the Maas at Nijmegen you turn upstream you can cruise that part of Holland which extends southwards between Belgium and Germany. This waterway is much used by large commercial craft and is not generally thought of as a cruising ground by the Dutch, so I will describe it only briefly. From Grave to Vise on the Belgian border is about 110 miles. Going upstream from Grave, the first part is canalised river passing through pleasant, often wooded, countryside. The first lock is just north of Afferden where new locks are being constructed. About 25 miles from Grave you come to Geijsteren where a watersports centre is being created in a couple of lakes on the eastern side of the river. A mile further upstream there is a jachthaven on the west bank at Wanssum.

At Venlo, a larger town 18 miles beyond, there are mooring places for yachts just before the twin rail and road bridges. Further upstream you come to the old locks and new double locks at Belfeld, with a recommended overnight mooring place on the downstream side of the waterfall. The large town of Roermond, just above the third lock, has a jachthaven in the Haven Bon Aventure, and another just to the south of the town at Roderbrug. The river is being straightened at this point and for some years to come construction traffic on the river is to be expected. Between Roermond and Maasbracht a bow has been taken out of the river and a new double lock is being constructed. From here to Maastricht the river reverts to its natural state and is equally shared by Belgium and the Netherlands, there being a joint agreement on navigation between the two countries. The river is navigable and quiet, but yachts normally lock into the Julianakanaal which rejoins the Maas just north of Maastricht. There are mooring places for yachts just to the south of the town. About 3 miles further to the south the Maas passes into Belgian territory and so out of the brief of this book. This part of Holland was one of the earliest to be inhabited, many towns dating back to Roman times. Most of those along the river contain much of architectural and historical interest. The local VVV offices can furnish all the details you may require. Most guide-books contain information about this area so you can decide before your cruise gets further than the preliminary stage if there is sufficient of interest in this area to make your visit worthwhile. This river and the Julianakanaal needs a fairly large motor cruiser if you are to cruise in comfort and safety.

Other canals

There is one branch of the river that is worth mentioning. Just north of Maasbracht is the Zuidwillemsvaart. This canal, with 19 locks, is smaller than the Maas and the Julianakanaal, taking craft of not more than 600 tons. Part in Belgium and part in Holland, it forms an important commercial alternative to the Maas. It is 75 miles long and joins the Maas again at 's-Hertogenbosch. The least headroom under the bridges is 12ft and there is a speed limit of about 6 knots for small craft. The Eindhovenskanaal leads off the Zuidwillemsvaart and goes to the industrial town of Eindhoven, the fifth largest city in the Netherlands. The growth and importance of Eindhoven began in 1891 when Gerard Philips started to manufacture incandescent lamps there with a staff of twenty-six.

Tilburg, which grew during the nineteenth and twentieth centuries, and Breda, a much older city often called The Hague of the south, can both be reached from the Zuid-willemsvaart by turning off into the Wilhelminakanaal, one of the few canals in the Netherlands to have staircase locks. The least headroom on this route is 15ft and the small craft speed limit is 6 knots. From Breda you can cruise through the Donge northwards and into the Amer at Willemsdorp. It is unwise at present to try to reach the Hollandsch Diep directly from Breda because of the construction work on the new canal. If you wish to travel westwards by canal from Breda you should obtain local advice to find out which canals are navigable at that particular moment.

THROUGH ROUTES

There are four main ports of entry into Holland from the North Sea. Vlissingen, the Hoek van Holland, Ijmuiden, and Den Helder. Of these Vlissingen is to be preferred if sailing from the south of England or from the Belgian or French coast. If sailing from the east coast of England the Hoek van Holland has much to recommend it as an entry port. In rough weather it is certainly easier to enter than is Vlissingen. From Germany, entry into Holland is normally made by crossing the estuary from Emden to Delfzijl, or by sailing down the Rijn to Arnhem. From Belgium you can enter Holland inland through the Maas, or from Antwerp you can enter the Westerschelde. From Ghent you can travel by canal to Terneuzen.

Within the Netherlands waterways system a number of waterways afford quick passage for craft from one part to another, although they may not, for lack of interesting places to see or density of industry, be canals normally thought of as suitable for cruising. I have listed below a number of these through routes which I have found to be satisfactory.

Zeeland–Rotterdam

Pass under the Haringvliet bridge into the Haringvliet. The headroom is about 46ft at low water and 42ft at high water. The southern end of the bridge has a headroom slightly lower than the northern end. Turn north into the Spui, then west into the Oude Maas and so enter the Nieuwe Maas. This route is preferable to the route through Dordrecht since the scenery is better, there is less commercial traffic and you reach the Nieuwe Maas on the same side of the railway bridge as the jachthaven. It took me three years to discover this route and I strongly recommend it. It is however tidal (as are all the other routes from Zeeland to Rotterdam) and the tidal streams are complicated. If you have a relatively slow boat it is well worth your while to spend a few minutes computing the state of the tides and the direction of flow at each place along the route at your estimated times of arrival at these places. Time so spent will be amply rewarded. To help you the ANWB publish tidal stream charts of the area.

There are no locks, so that delays are likely to occur only to yachts with pole masts above 30ft and then only at the lifting bridges in the Oude Maas. There are no severe speed restrictions. The only warning note to sound is that the entrance channel to the Spui from the Haringvliet is twisting and although very well buoyed it is tempting to try cutting the corner. My advice is not to try it, particularly on a falling tide, unless you have the best part of six hours to spare. If you do go aground on a falling tide and have to wait for the flood to refloat your craft it will put your tidal calculations out and so you will probably have to complete the trip against a foul tide. The area is predominantly rural with no towns of much interest.

Rotterdam–Amsterdam

The only route left for yachts with fixed masts is through Gouda and the Braassemermeer. Even this route has many lifting bridges and delays can be expected. The bridges over the Gouwe and at Alphen a/d Rijn have a lowered headroom of 8ft and do not open at all on Sundays. The last bridge in Amsterdam before reaching the Noordzeekanaal on this route is a railway bridge with headroom of 16ft, which only opens at night for a short time around 01.00 hrs. It is almost impossible to do this trip in a day with a fixed mast yacht, and if you are unlucky it could take three.

If your ship's headroom is less than 14ft, a better route is via the Gouwsluis and then the Uithoorn so as to enter Amsterdam along the beautiful Amstel River. This route is not only faster but also has the better scenery. Powerful motor cruisers, especially on Sundays, would find the longest route the quickest, that is via the Lek to Vreeswijk, then by the Lekkanaal and the Amsterdam–Rijnkanaal.

Amsterdam–Den Helder

This is directly through the Noordhollandskanaal. The trip can be easily completed in a day even by a slow sailing boat with a fixed mast and a small auxiliary. The skipper of a high speed motor cruiser may like to air his cylinders, in which case he should lock out into the Ijsselmeer where there are no speed restrictions and then lock out again into the Waddenzee and enter Den Helder from the sea.

Amsterdam–Friesland

With the construction work on hand in the Ijsselmeer it is now better to take the southern route and to follow the barges along the north shore of Flevoland before turning north to follow the shore of the Noordoost polder to Lemmer. This way you will have nothing to see except the waters of the meer on your left and the walls of the reclaimed polders on your right. At present there are locks only at the ends of this route but soon there will be one in the middle as well.

Lemmer–Delfzijl

The quickest route is suitable for yachts with a headroom of less than 22ft. It goes via Grote Brekken, Sneekermeer, Pikmeer, Fonejacht, Bergummermeer, Kaper Roblisdiep, van Starkenborghkanaal and the Eemskanaal. There are three locks and the distance is about 75 miles. There are speed restrictions of about 5 knots.

For sailing yachts with fixed masts and a draught of less than 6ft the route described above is possible except that fixed mast yachts will have to pass through the town of Groningen instead of using the van Starkenborghkanaal. The bridges in Groningen do not open on Saturdays or Sundays. The only route for deep draught yachts, that is up to 6ft 3in, runs from Lemmer to Fonejacht as above and then to Leeuwarden and Dokkum and into the Lauwersmeer. It then runs along the Reitdiep to Groningen before joining the Eemskanaal. Many bridges in the province of Groningen do not open at all on Sundays and some of these have headroom as low as 3ft.

WATERWAYS NOT RECOMMENDED

Criteria for excluding certain waterways from a list of cruising grounds may be summarised as follows. Firstly, density of commercial traffic. There is little to be gained, other than grey hairs, from dicing about amongst enormous ships in a small craft, though many of the larger motor cruisers may be able to use very commercialised waters with safety. It is as well to remember, though, that the presence of a yacht may impede the navigation of some commercial craft in a particular water and since the skipper of the barge is earning his living, whilst the yachtmaster is simply enjoying himself, it is perhaps more considerate to leave such waters alone. Secondly, there are some areas where a great deal of construction work is in hand. Leave the canal builders to get on with their job and postpone your visit until the work is completed. Thirdly, the presence of heavy or smelly industry can create areas not suitable for cruising, which can injure both your health and your craft. Lastly, there are areas where navigation is difficult. Remember that dangers in such areas can range from a few embarrassing hours stuck on the mud to the loss of your ship and possibly your life as well.

Of course the importance of these factors is dependent upon the type of craft you are using and the experience of your crew. With regard to the commercial areas and construction works what is undesirable on a weekday may become a useful and interesting cruise ground at the weekend. What follows therefore should be taken in the general context of your own experience, your choice of boat, the prevailing weather and even the day of the week.

Starting with the sea area, the sandbanks off the Oosterschelde together with the mixture of tidal flows there make that area undesirable as a cruising ground except in very settled weather conditions. Any yacht skipper proceeding seawards of Zierikzee should exercise the greatest caution. The same conditions make the entrance into the Waddenzee from the North Sea between the mainland and Texel a dangerous place. The other entrances to Holland, Vlissingen, Hoek van Holland and Ijmuiden, although they carry a large amount of commercial traffic, are probably far less dangerous. A supertanker is somewhat easier to see than is a bank of hard-packed sand a foot or so under the water in a tide race.

The Amsterdam–Rijnkanaal, the Zuidbevelandkanaal and the Lekkanaal should be avoided as they carry a great deal of commercial traffic and there is an alternative route, quieter and more suitable, to each of these canals. The Noordzeekanaal and the Nieuwe Waterweg also have many large boats on them but sometimes have to be used by yachts as there are no suitable alternative routes.

The smaller canals of Amsterdam are also best left to the locals. Many are used by the canal buses that take sightseers around the waterways of the city. These are large in comparison with the water they have to manoeuvre in and if you want to visit these canals, take a trip on one of these purpose-built craft.

Much of the industry of the Netherlands is centred around Europort. You may come across pockets of industry in other places, but they are soon passed, so with the exception of the Nieuwe Waterweg and the Brielsemeer, there are no areas of any size where the water, or the air, is so polluted as to make it undesirable for cruising.

There are few mandatory controls on pleasure craft, but some canals and lakes are banned for motor vessels and some are banned to sailing vessels using auxiliary motors. These are only on minor waterways and they are clearly marked on the ANWB charts. In some places waters are reserved for canoes and in others for rowing boats. These are also marked on the ANWB charts.

CHAPTER FOUR

Living on the Canals

CLIMATE

The Dutch enjoy or suffer, depending on your point of view, a temperate maritime climate like the British. That is to say you can have three weeks' baking hot dry weather in the summer, or alternatively three weeks almost solid rain. For the most part in Holland you will get something in between.

SHOPS

Shopping in Holland presents few difficulties for English people. Dutch food is similar to English, but there is rather more of it. For example, I have found that a packaged meal from a Dutch store that states it is for two people will often satisfy four teenage boys. Dutch shops tend to specialise more than English ones and there are fewer supermarkets. It is no good being in a hurry in a Dutch shop. The assistant will serve one customer at a time. Most assistants speak English, know their jobs and also the products they are selling.

If you are cruising with children it is worth remembering that the Dutch are excellent pastry cooks. In many pastry shops and confectioners, children are given a sweet as a matter of course, so be warned if you are worried about your children's dental health.

Tobacco for the adults is cheap if it is Dutch, but English

and American tobacco, cigarettes and cigars are expensive. Dutch cigars and tobacco are very mild, but the tobacco is often scented. The size of the bowls of Dutch pipes is a reflection of this cheapness and mildness.

If you want to bring home something really Dutch, apart from a bottle of Oude Genever, then there is a lot to be said for a pair of clogs. They are excellent wear for gardeners. Bought from a proper clog shop and not a tourist boutique they are not expensive. Most clog sellers have a conversion table relating English sizes to clog sizes.

EATING OUT

Eating out in Holland presents no problems. In general, prices are in proportion to the quantity and quality of the meal. Certain restaurants carry a plaque showing a fork with a hat on carrying a camera and the words *Tourist Menue*. Such restaurants serve a tourist menu at a reasonable price. Many restaurants offer a *Kindermenue* designed for children. In general these are very acceptable to youngsters. The Dutch are not prone to *haute cuisine*, but serve a substantial meal, well cooked and efficiently presented. You should not tip in a Dutch restaurant as it may offend. The service charge and the value added tax will both be put on the bill.

A Dutch breakfast consists of coffee or tea, cereals, cheese, ham or sausage, a boiled egg, toast and marmalade. This is more in line with the English idea of starting the day than the general continental breakfast. The Dutch usually have a snack lunch and eat heartily in the evening.

EVENING ENTERTAINMENT

Information about evening entertainment in any town can best be found by contacting the local VVV office which will furnish you free of charge with a list of places worth seeing, museums, art galleries, shows, and what's on at the local cinema. Holland is a small country with a language that is not widely spoken outside its borders and so films are almost always of foreign origin and are shown in their original language with Dutch sub-titles. They are never dubbed in Dutch. This is convenient for English-speaking tourists if a little awkward for the Dutch.

Midget (or crazy) golf, often floodlit, is almost a national game and is excellent fun for adults, and children alike. Fairs are frequent and their prices reasonable. I am assured by my crews that those fairground devices designed to shake you limb from limb are more exciting than their English counterparts.

OTHER CANAL USERS

The non-cruising folk who use the inland waterways of Holland can be divided into two classes, the professionals and those using the waters for recreation.

Barges and bargees

Chief among the professionals are the bargees. Many Dutch barge skippers own their own barges and magnificent craft they are too. The skipper lives aboard with his family and it is common on the larger barges to take the family car on board. Young children are safely confined to a playpen on deck and the skipper and his wife share the navigation. When a barge is being moored it is common practice for the skipper to hand over the wheel to his wife while he attends to the more strenuous job of handling the mooring warps. A lead that some yacht skippers might follow with advantage.

Existing in a largely free enterprise world, the Dutch barge skipper cannot afford delays and inefficiency and he often has a tight schedule to meet. If you remember this and give way a little it will be much appreciated. Remember that for you, on a holiday cruise, a few minutes' delay here and there are of no consequence.

Trawlers

Second in number to the barges are the trawlers. Many old fishing harbours, although not now directly connected to the sea, still have their fishing fleets which now have to sail further to the fishing grounds than in former times. The trawler is a fast, highly manoeuvrable little ship often manned by a minimum crew. The skill of the men who handle these boats never ceases to amaze me. In most harbours that are still used for trawlers and which now have berths for yachts as well, each have their respective areas. If you keep to the yacht berths you will be safer and it may well surprise you to find that a harbour that was practically empty when you turned in for the night contains up to twenty trawlers in the morning, and all without waking you.

Other professionals

These include the men who build the waterways and make them work. Dyke building is very skilled, dangerous and strenuous. The Dutch are past masters at it and have a great deal of specialised equipment for the job. It is well to keep clear of such works as they are rather like icebergs, in that there may be dangers under the water some considerable distance from the dredgers, pile drivers, barges and other craft that you can see on the surface.

The officials who make the waterways work, of whom the lock and bridge-keepers are the two most often met by the cruiser folk, have a very important job to do and do it with courtesy and a total lack of officiousness. The Rijkswaterstaat and the Rijkspolitie control the waters and those who use them. They have excellent well-equipped craft that can cope with most waterway emergencies. Orders from them must be obeyed without hesitation.

Fishermen, water-skiers, and swimmers

Recreational users other than the cruising folk are chiefly fishermen, water-skiers and swimmers. Water-skiing in

Holland is restricted to certain areas and the Dutch seldom ski in forbidden waters. Fishing is a popular pastime, but the fishermen have not the enmity towards other waterway users that they seem to have in some countries. The coarse fishing in Holland is good and a licence can be obtained by visitors without difficulty. You should apply at a principal police station.

Swimming, especially among children, is very popular and since the level of pollution in the canals and lakes is low you may expect to find swimmers almost anywhere. Official swimming places are marked on the ANWB charts and you should be especially careful, particularly if under power, when passing these places. It is advisable for visitors to seek local advice before swimming unless you swim at one of the designated *Swimbad.*

Unseen users

All over the shallow parts of the lakes, fishing stakes can be seen. In four years I have only once seen anyone attending these fish traps, but they are obviously very much in use. They form at one and the same time a danger to navigation and a livelihood for their owners. It is as well to avoid such areas. A fish trap wound round the propeller of your boat is unlikely to improve the boat's performance and is certain to upset the owner of the fish trap.

Ducks and mosquitoes

Ducks abound in Holland and are prone to set up incessant quacking at the slightest provocation, usually in the middle of the night. They swim about all over the canals, lakes and rivers. There is no point in taking avoiding action. The ducks always manage to get out of the way even if their dignity is somewhat upset. You may like feeding ducks in the park at home, but in a Dutch jachthaven this is not a pastime I would recommend as your boat will then be surrounded by quacking birds long after you wish to turn in.

Waterways attract mosquitoes and although the young eels eat the larvae, some mosquitoes survive to maturity and you should equip your boat with mosquito netting to cover any hatchways that are left open in the evening. If you are particularly prone to mosquito bites, than a few pence spent on some insect repellant could be a good investment.

REPAIRS

If your craft should break down, or you should be involved in some accident and need repairs to damage, these will not in general present a serious problem. If your craft is one hired in Holland, you should consult the owners. If, however, it is your own boat then the nearest yacht club or jachthaven can advise you where to obtain the spares you want or where to get the repairs done that you need. In general the prices for repair work are higher than in England, but the service is more rapid and often of a much higher quality. Remember, however, that if your trouble is mechanical and the machine is British you may have trouble getting spares. In general spares obtained from England are subject to import duty in Holland and may be held at the central post office in 's-Gravenhage until you pay the duty. Running repairs done to your boat outside England are not subject to duty when you return, but you should declare what repairs have been done. This rule only applies to repairs and does not include modifications or additions to your boat. These are both subject to import duty on returning to UK.

There are plenty of ship's chandlers in Holland. Yacht clubs can always advise on the nearest and best. They normally carry an extensive stock so that minor replacements to the boat's gear as distinct from engine replacements can usually be achieved with little trouble.

HOUSEBOATS

Land is scarce in Holland and in consequence there are a

larger number of houseboats in proportion to population in Holland than almost anywhere else in the world. The Dutch idea of a houseboat is rather different from the English. The houseboat in the Netherlands is normally a floating platform on which a house or bungalow is built. Some are quite expensive and some are very picturesque. Because they float they are susceptible to wash, so exercise consideration when passing them. It is very tempting to peer in at a houseboat's windows, but one should remember that it is somebody's house and curb one's natural curiosity.

FUEL

The obtaining of fuel presents no problem even if you run out during the day as roads run alongside most of the canals and on these roads there are garages to supply your need. Many will have small landing stages to tie up to. It may be more convenient to fill up at a jachthaven where you can fill your tank directly, but it is worth carrying a spare can for fuel in case you have to get fresh supplies from a roadside garage where you may have to carry the fuel across the road.

CYCLE HIRE

Most towns of any size have at least one cycle-hire agent. The prices for these machines are very reasonable and enable cruising folk to do a little exploring away from the actual waters. Certainly they provide the best way of seeing the Friesian islands. The cycles are of a standard pattern and can be adjusted to suit most sizes of rider. The Dutch cycle is a little different from the British machine in that there is normally only a front hand-operated brake. You operate the rear brake by pedalling backwards. This does not take long to get used to but can be embarrassing as you ride away from the hire agent if you haven't been warned first. Remember to ride on the right-hand side of the road and to use the cycle tracks that are normally provided.

Page 107 (*above*) The old and the new. Scene on the River Ijssel; (*below*) motor boats in the jachthaven at Zutphen

Page 108 (*above*) Entering Amsterdam from the south; (*below*) ferry in Amsterdam

PUBLIC TRANSPORT

Public transport is good in Holland. The train service is excellent and by no means expensive. Buses are frequent, clean and fast. The major towns still have excellent tram services. In general buses and trams have fixed fares and you can either buy a book of tickets before boarding, or pay the driver. It is forbidden to smoke or eat, even an ice-cream, on a Dutch bus or tram. Taxis are available in Holland but they are far more expensive than in England.

CHAPTER FIVE

Navigation

BRIDGES AND LOCKS

Because Holland is a flat country, many bridges have very limited headroom. Most important roads and railways, however, are built up as they approach canals so that they allow sufficient headroom for most motor cruisers and commercial craft. Less important roads, however, cross the canals at lifting bridges, the opening times of which vary from district to district and also depend on the density of traffic. Each year the ANWB publish lists of the opening times of all bridges, both road and rail, that have regular opening times. Unlisted bridges will open on request during certain times that normally coincide with average daylight hours, often with some restrictions on Sunday openings. A rough general rule is that bridges, with the exception of motorway and railway bridges, open frequently during the day-time except during the morning and evening rush hours, and in some towns and in the province of Friesland during the midday lunch period. The length of the rush hour periods is roughly proportional to the size of the town in which the bridge is situated, or the town that the road serves. The opening times are always displayed in the bridge-keeper's office window and so if you are delayed you can tie up and go and find out when the bridge will open next.

Types of bridge

There are three main types of lifting bridge and a few other rarer kinds. The cantilever is the commonest, where the counterbalanced roadway swings up into an almost vertical position, thus giving unlimited headroom for sailing yachts. Such bridges, however, have no centre island and so waterborne traffic in both directions has to pass through the same lifting span. It is therefore essential to obey the signals, detailed later, in order that you may avoid risk of collision with traffic coming in the opposite direction. Nearly all such bridges are power-operated and if you are the only craft the bridge-keeper may sometimes only open the bridge sufficiently for your boat to pass.

The next most common type of opening bridge is the swing bridge. This pivots about a central island. There are three operations involved in opening this type of bridge apart from stopping the traffic. First the supports have to be lowered. Secondly the bridge-keeper has to move to the centre of the bridge and swing the bridge so that it is in line with the stream. Thirdly he has to operate the signals. These bridges are mainly hand-operated and because of the central island traffic can often pass in both directions at the same time. However you should watch for the signals, as sometimes navigation is not possible through one half of the opening.

The third type consists of a span lifting vertically employing a lot of cables, pulleys and counterweights. These bridges still have limited headroom when raised but it is always of such a height that it need cause no concern to yachts. Bridges of this type are often only raised far enough for the yacht to pass underneath. I have never known a bridge-keeper miscalculate, thought from my distorted viewpoint on board I have often had doubts. I suggest you proceed with caution through these bridges, although not too slowly as they are often rail bridges with limited opening times. This system is ideal for railways because the overhead power cables rise with the bridge.

Other types of bridge include floating ones that are withdrawn into the bank, swing bridges that pivot on one side of the canal, and cantilever bridges that open on both sides and meet in the middle when closed. All bridges are operated for you. Normally there are places to tie up on each side of the bridge and where the opening times are restricted there are also catwalks to enable you to get ashore. Bridge signals are the same as the general lock signals, they are as follows:

Bridge (lock) Signals

Two red lights	Bridge (or lock) not due to open. Delays can be expected
One red light	Bridge (or lock) opening soon or already open. Passage prohibited
One green light	Bridge (or lock) open. Passage permitted

To signify to the bridge- (or lock-) keeper that you wish to pass you should sound three long loud blasts on your ship's horn. In most cases the keeper will answer by extinguishing one of the red lights. If this does not happen it is either because he has not heard you, in which case you should try again nearer his office, or you have arrived at one of the times when the bridge or lock does not open. If the latter is the case then threats, bribes, general cajoling or repeated blasts on your horn will delay rather than speed your passage. If you have to wait it is best to tie up and make tea.

Payment and formalities

Payment and formalities on passing locks and bridges is something of a mystery to me since there seem to be no general rules, and although the locals seem to know what to do intuitively, visitors are often mystified. I have adopted the following stance and this seems to work. Firstly I assume all bridges are free unless someone lowers a receptacle on a string, in which case I put a *kwartje* (25 cent piece) in it. All locks I assume to be free unless I see people from other boats going towards an office, or I see a man in uniform obviously expecting some payment, in which case I pick up some small change and hop ashore. Where there is a charge for the use of a lock, you may have to answer a few questions such as: the name of your yacht; where you have come from; where you are bound for; and either the length of your boat in metres, or the tonnage of your boat. The answers will duly be recorded in the lock-keeper's log book and on the receipt you will get for the money you pay. The charges are also a mystery since they are often as little as 10 cents (a little over one new penny).

BUOYS

The Netherlands Government uses the *Uniform International System* of buoyage, which combines the *cardinal* and *lateral* systems. The cardinal system is used at sea and the lateral system in estuaries, harbour approaches, lakes, rivers and canals. Drawings of the lateral system buoys are to be found on page 114.

Working dredgers are often surrounded by a number of buoys marking the position of the moorings of the vessel. Dredgers use these moorings to pull themselves over the area that they are currently working, and so such buoys may well be clear of the dredger. Dredger buoys lead a hard life and even if they were painted once, the colours are more often than not totally obscured by rust. The only safe rule is to keep well away from dredgers at all times and to avoid trying to make passage between any buoys in the vicinity of the dredger and the dredger itself. Do not be misled by barges which may well be going to the dredger and the skippers of which know the precise layout of the mooring system.

Buoys in general

At the risk of being obvious, I will point out a few general facts about buoys that are often overlooked in books and chapters on navigation because they seem self-evident to the authors. A buoy is moored to the bottom by a chain and at all times, but particularly at low water, there is far more chain than the depth of water. When the wind and tide are

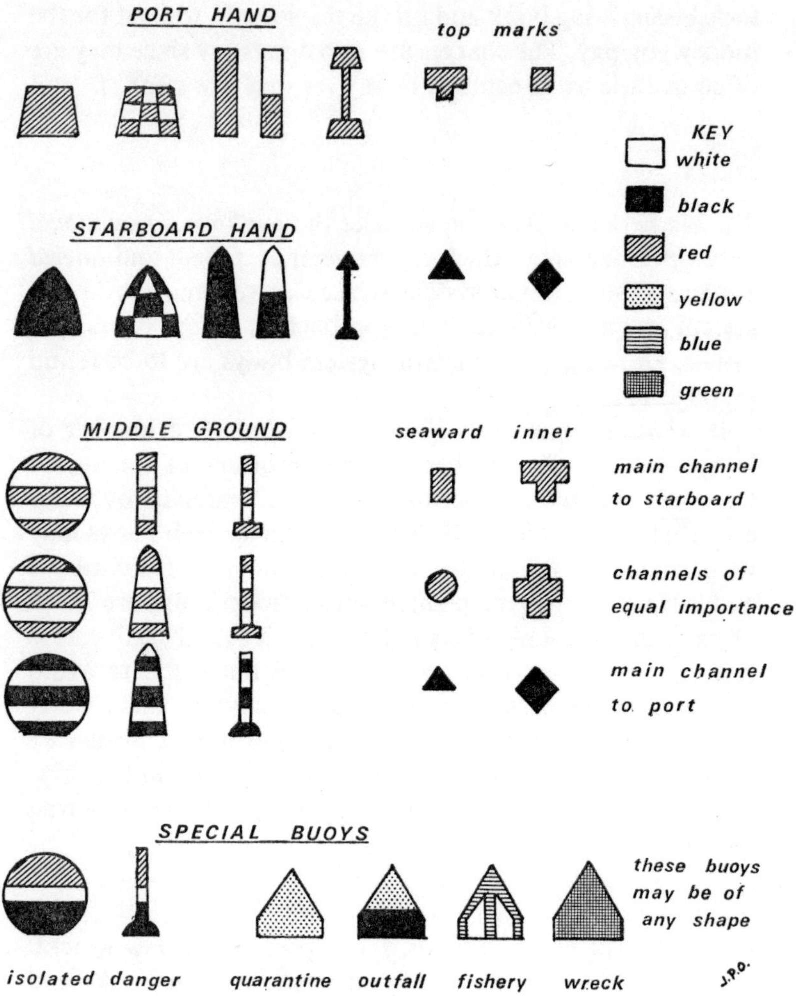

Buoys to be found on the inland waterways

complementary the buoy may have drifted a fair distance from its station and so the visible marker is far from over the top of the sinker. Thus in a narrow channel with a cross wind the whole of the buoyage will be drifted towards the downwind side. Where the banks of the channel are fairly gradual this is of little significance but if the banks are steep to, then you could well be in some danger if you are following the lee side of the channel.

From time to time buoys become damaged and are taken off station so you may find an occasional buoy missing. Buoys are sometimes moved, especially in waters like those of Holland where the channels are constantly changing, so you should be prepared for the fact that the buoys may not be positioned exactly as shown on the chart. It is always unwise to place reliance on one single buoy for fixing your position. It is better to confirm your position by sighting at least one, and preferably two, other buoys.

BEACONS

Where the rise and fall of water is small and the depth of water not more than a fathom or two, beacons are often used in place of buoys, particularly on minor waterways. Beacons have the advantages that they are not blown about by the wind or drifted by the current and are cheaper to maintain. Beacons can also be placed on shore and are particularly useful in showing the direction of the channels in rivers, estuaries and creeks. In Dutch waters there are four main types of beacon: lateral beacons, perches, *raambaken* and *bolbaken*. These beacons are drawn on page 116.

Lateral System Beacons

These are similar to spar buoys, except that they are driven into the bed of the waterway instead of being carried on floating buoys.

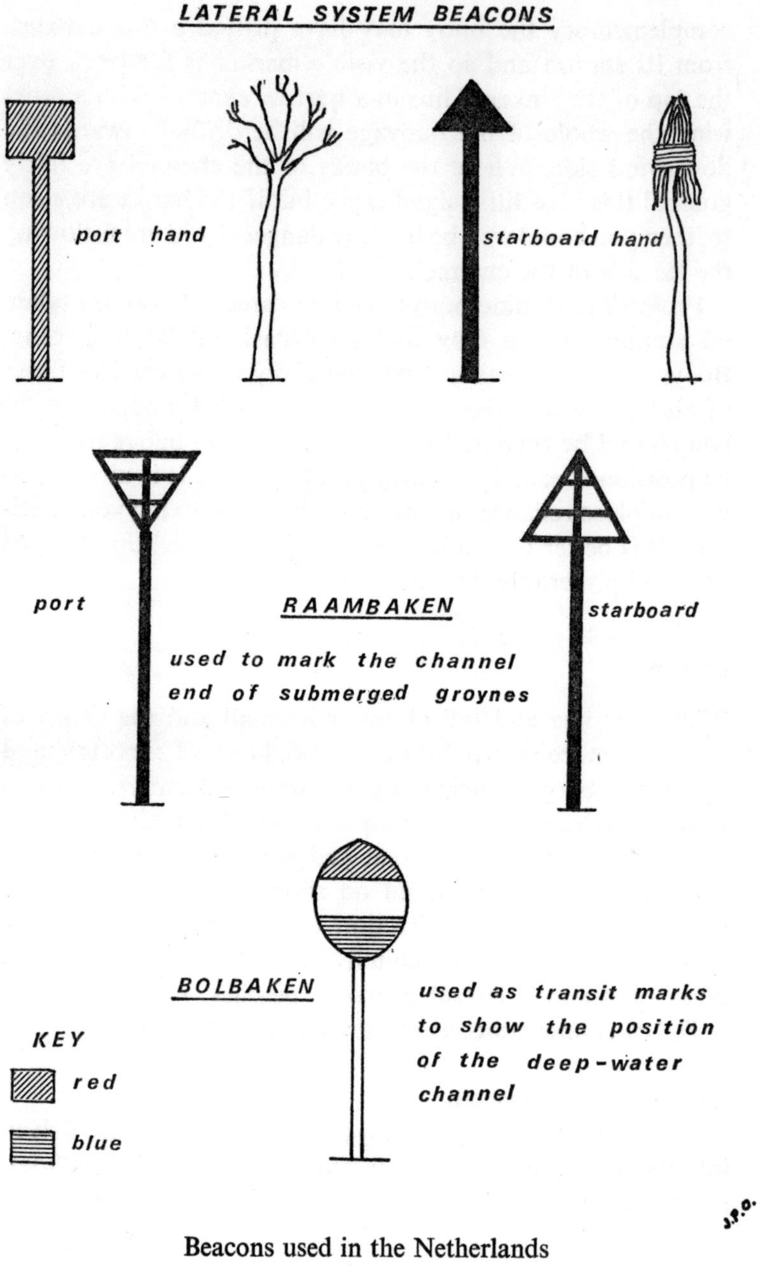

Beacons used in the Netherlands

Page 117 (*above*) *Puffdragon* tied up by the roadside at Makkum; (*below*) houseboat on the River Vecht

Page 118 (*above*) A dredger at work; (*below*) quite a small barge by Dutch standards

Page 119 (*above*) A typical cantilever bridge; (*below*) passing through a swing bridge

Page 120 (*above*) A *tjackshalt*. A diesel-powered reproduction of an old horse-drawn canal boat; (*below*) Radboud Castle

Raambaken

In the great rivers where there are many groynes to deflect and tame the waters, the ends of these obstructions are often marked with frame beacons known as *raambaken.*

Bolbaken

These beacons are used extensively on rivers to indicate the position of the deep water channel. They are placed on the bank of the river when the deep water channel does not run in the middle and are sited on the bank nearest the deep water. When the channel changes from one side of the river to the other a pair of *bolbaken* are used, the landward one being taller. The *bolbaken* can thus be used as transit posts.

OTHER MARKERS

Kilometre posts

On the great rivers and on some of the canals in addition to the buoys and beacons mentioned above there are also kilometre posts. These are markers placed every kilometre along the length of the river or canal and are white rectangles with black numbers. They are useful for checking your speed and in the more rural areas can give you an accurate fix of your position along the particular stretch of water you are sailing. These kilometre posts are marked on the ANWB charts. In the great rivers the numbers indicate the distance in kilometres from the centre of the Rijnbrug at Konstanz, but with some of the bends in the river now cut off the distances shown overall are a few kilometres greater than the actual distance that would have to be travelled.

Service markers

Finally, along the banks of canals and rivers you will see square yellow plates with a black border and a single letter in black. These indicate where services such as underwater cables, gas pipes, etc, cross the waterway. For obvious reasons it is forbidden to anchor near such crossings. There are no other navigational signs that should not be either obvious or familiar to anyone who has had experience of cruising in Britain or Europe. If, however, you should meet a sign whose significance you do not understand then be prudent and keep away, or if you have to go near it, proceed with caution.

CHAPTER SIX

Canal Regulations

There are three basic sets of regulations governing the conduct of craft on the waterways of the Netherlands; the *Vaarreglement*, which are the general rules for preventing collisions; the *Algemeen Reglement van Politie voor Rivieren en Rijkskanaalen*, the police regulations for the rivers and canals; and the *Reglement van Politie voor Rijnvaart met het Reglement van Politie voor der Merwede en de Noord*, the special regulations for the Merwede, Rijn and Noord. Unless you are cruising these last three rivers you can ignore the special regulations that apply to them.

Basic rules

Three basic rules override anything contained in the regulations. Firstly, as a small ship the cruiser has to give way to commercial craft. Secondly, the onus is on the master of any and every ship to avoid a collision. Thirdly, the master of a ship is required to disregard the regulations in special circumstances, if in the interests of good seamanship it is necessary for him to do so.

VAARREGLEMENT

1. The regulations regarding lights, sound signals and the avoidance of collisions are substantially the same as those contained in the *International Regulations for preventing Collisions at Sea.*

2. A vessel travelling upstream may determine the course of a vessel travelling downstream by showing *by day* a light blue flag, or *by night* a flashing white light on her starboard side. Passing will then take place starboard to starboard. On all other occasions passing must be port to port. Yachts will not give these signals but are expected to obey them.
3. A vessel must normally overtake on the port side. A vessel wishing to overtake on the starboard side will sound two long blasts and one short blast. This signal is also given by a vessel which is about to be overtaken but wishes the overtaking vessel to pass on the starboard side.
4. On each vessel under way, the helm must be handled by a capable person at least sixteen years old. Small sailing boats with *no* auxiliary power are the only exceptions.
5. A copy of the *Vaarreglement* must be carried aboard every vessel at all times. (See section on formalities in Chapter Two.)

ALGEMEEN REGLEMENT VAN POLITIE

1. A copy of the *Algemeen Reglement van Politie* must be carried aboard every vessel at all times. (See section on formalities in Chapter Two.)
2. A vessel is only regarded as a sailing vessel when under sail and using no mechanical power.
3. It is forbidden to enter a harbour when a red flag is flying by day, or a second light is showing next to the red harbour light at night.
4. The opening of a lock for sluicing is notified at least one hour before by a blue flag with '*Spuien*' in white letters by day, or three red lights at the corners of a triangle at night.
5. Vessels must reduce speed when nearing locks and bridges. A red flag, a red board, one or more red lights or a red cylinder topped by a red hemisphere indicate that passage through a lock or bridge is prohibited.

6. Sails must be struck when passing through bridges and locks. Downstream-going vessels have priority at bridges. Where there is no current, vessels going before the wind have priority.
7. When passing vessels, speed must be reduced so that there is no danger from wash.
8. Vessels moored along the towpath must help pass tow-lines.
9. Captains of stationary vessels must shift in harbours and canals if ordered by Department of Works officials. Wherever possible stationary vessels must be tied to moorings or dolphins.
10. Captains of vessels moored at landing stages must allow other vessels to tie up alongside and share access to the shore.
11. It is forbidden to operate bridges, locks and other types of waterway works.

REGLEMENT VAN POLITIE VOOR DE RIJNVAART

1. The rules regarding lights and general conduct in order to avoid collisions are the same as in the *Vaarreglement*.
2. Pollution of the water with oil is forbidden.
3. In the case of accidents involving risk of loss of life or obstruction of the fairway, every captain must render all possible assistance compatible with the safety of his ship.
4. The regulations regarding the meeting of two vessels where there is a current are the same as in the *Vaarreglement*.
5. It is forbidden; to come alongside a moving vessel without the permission of the captain; to swim or water-ski near moving vessels.
6. Speed must be reduced so as not to cause wash or suction that may cause damage.
7. Passage under fixed bridges is prohibited when the span is marked with either: one or two red lights; one or two

red flags; or a red board with a white horizontal stripe. The same signals are used for weirs.

8. It is forbidden to use engines in locks from the time the vessel is moored to the time it is ready to leave.
9. Sound signals are not given by yachts but are included here so that cruising yachtsmen may interpret the actions of commercial craft and thus not find themselves in a position of danger. The commercial craft skippers will assume that the yachtsman understands the signals and will take appropriate action.

(i) Vessels which hear the signal consisting of a repeated blast of three tones of ascending key must

(*a*) If they are close to the bank, keep close and if necessary stop until passing has taken place, or

(*b*) If they are not close to the bank, clear the channel as quickly as possible.

The three-tone signal is only given by vessels navigating by radar which have seen another vessel on the radar screen whose position and behaviour could lead to danger.

(ii) Other sound signals.

(*a*) *General signals*

Signal	*Meaning*
one long blast	attention
one short blast	I am going to starboard
two short blasts	I am going to port
three short blasts	my engines are going astern
four short blasts	I cannot manoeuvre
series of very short blasts	there is danger of collision
repeated long blasts or series of bell strokes	emergency signal

(*b*) *Signals given by two vessels meeting*

For intention to pass port side to port side

Normal event

One short blast from vessel going upriver	I wish to pass port side to port side

Signal	*Meaning*
One short blast from vessel going downriver	You can pass port side to port side
Deviation	
Two short blasts from vessel going downriver	No, you MUST pass starboard to starboard
Two short blasts from vessel going upriver	I will pass starboard to starboard
For intention to pass starboard to starboard	
Normal event	
Two short blasts from vessel going upriver	I wish to pass starboard to starboard
Two short blasts from vessel going downriver	You may pass starboard to starboard
Deviation	
One short blast from vessel going downriver	No, you MUST pass port side to port side
One short blast from vessel going upriver	I will past port side to port side

(*c*) *Signals when overtaking*

For intention to overtake a vessel on her port side	
Normal event	
Two short blasts followed by two long blasts from the overtaking vessel	I wish to pass you on your port side
No sound signal from the vessel being overtaken	You may pass me on my port side
Deviation	
Two short blasts from the vessel being overtaken	No, you must pass me on my starboard side
One short blast from the overtaking vessel	I will pass you on your starboard side
Intention to overtake on the starboard side of a vessel	
Normal event	
Two long blasts followed by one short blast from the overtaking vessel	I wish to pass you on your starboard side
No sound signal from the vessel being overtaken	You can pass me on my starboard side
Deviation	
One short blast from the vessel being overtaken	No, you MUST pass me on my port side
Two short blasts from the overtaking vessel	I will pass you on your port side

(*d*) *Signals when entering or leaving harbours and secondary channels*

Signal	*Meaning*
Three long followed by one short blast	I am going to starboard
Three long followed by two short blasts	I am going to port
Three long blasts	I am going to cross the channel

(*e*) *Fog signals*

One long blast repeated at least every minute	Given by a single moving vessel
Two long blasts repeated at least every minute	Tug and tow, or push-tow moving
Three times three tones of different key. Successive and without interruption	Signal given by vessels going downriver in fog with the aid of radar
One series of bell strokes at least every minute	I am stopped on the port side of the channel
Two series of bell strokes at least every minute	I am stopped on the starboard side of the channel
Three series of bell strokes at least every minute	I am stopped, but do not know on which side of the channel

CHAPTER SEVEN

Other Information

CHARTS AND BOOKS

Charts

Due to the constantly changing pattern of the sandbanks in the estuaries and the speed with which works are carried out on the inland waterways it is essential to use the most up-to-date and detailed charts. For open waters the best are undoubtedly the Dutch small boat charts (*Kaart voor Zeil en Motorjachten*). These are published by the Netherlands hydrographic service and are reprinted every spring. Those of particular interest are:

1803 Westerschelde
1805 Oosterschelde en Veersemeer
1807 Grevelingen, Krammer, Volkerak, Haringvliet en Hollandsch Diep
1809 Nieuwe en Oude Maas, Dordtsche Kil, Amer en Brielsemeer
1810 Ijsselmeer
1811 Waddenzee (Westblad)
1812 Waddenzee (Oostblad)

These charts are coloured, detailed and contain all the information needed for navigation by a cruising yacht, including the position of jachthavens and the facilities provided at all the major harbours. The legend and all other important information is printed in both Dutch and English. The charts can be obtained from:

ANWB Hoofdkantoor, Wassenaarseweg 220, 's-Gravenhage

J. Datema, Nautisch en Technisch Bureau 'Atlas', Ged Zuiderdiep 15, Groningen

L. J. Harri, Prins Hendrikkade 90, Amsterdam

N. V. Observator, Wasteland 30, Rotterdam

Fa. Datema, Waterstraat 1, Delfzijl

For the canals and rivers the ANWB Waterkaarts should be used. There are sixteen of these charts as follows:

A	Midden Friesland	J	Grote rivieren westblad
B	Zuidwest Friesland	K	Grote rivieren middenblad
C	Noordwest Overijssel	L	Grote rivieren oostblad
D	Gelderse Ijssel	M	Limburgse Maas
E	Randmeren	N	Biesbos
G	Amsterdam Alkmaar	O	Veersemeer
H	Hollandse plassen	P	Vinkeveense plassen
I	Vechtplassen	R	Loosdrechtse plassen

These charts can only be obtained from ANWB offices, at any of the following towns:

Alkmaar, Amersfoort, Amstelveen, Amsterdam, Apeldoorn, Arnhem, Assen, Breda, Dordrecht, Eindhoven, Emmen, Enschede, 's-Gravenhage, Groningen, Haarlem, Heerlen, 's-Hertogenbosch, Hilversum, Hoogvliet, Leeuwarden, Maastricht, Middelburg, Nijmegen, Rotterdam, Schiedam, Terneuzen, Tilburg, Utrecht, Ijmuiden, Zaandam, Zwoole.

They are open on weekdays from 08.45 until 16.45 and on Saturdays from 08.45 until midday.

If you wish to obtain the charts by post, then you should address your enquiries either to ANWB Kantor Museumplein 5, Amsterdam, or ANWB Hoofdkantoor, Wassenaarseweg 220, 's-Gravenhage, enquiring the current price and the cost of postage and enclosing an International reply coupon obtainable (price 10p) at any main post office.

Other ANWB publications

The ANWB produce and stock a large number of publications of interest to anyone touring Holland by whatever means. Of particular interest to yachtsmen cruising the waters of the Netherlands are:

Openingstijden van Spoorbruggen. This gives the opening times of all lifting railway bridges.

Zondagsbediening van Bruggen en Sluicen. This gives the opening times of all bridges and locks on Sundays. It consists of several pamphlets for the different areas.

A Short Description of the Cruising Possibilities in Dutch Waters.

Verhuurders van Motorjachten. This is a list of recommended hirers of motor jachts. It is written in Dutch.

Verhuurders van Zeiljachten. This is a list of recommended hirers of sailing yachts. It is written in Dutch.

Almanak voor Watertoerism. This is in two parts (both in Dutch). Part One contains a great deal of information including the canal regulations, *a copy of which you are required to have aboard at all times.* The fact that they are in Dutch and you may well not understand a word of them is neither here nor there in the eyes of the law.

Tidal stream charts

The Netherlands Hydrographic service produce *Stroomkarten,* to help you determine the direction of the current at any particular time. They consist of a chart with small holes in it at various places on the waterway and a larger hole in one corner. Inside the chart is a card which you pull out until the number of hours before or after high water at some reference port shows in the larger hole. Arrows in the smaller holes in the chart then show the direction of the current at those places. This may sound like playing childish games but in the estuaries nothing short of a mini-computer will enable you to calculate the tidal flow as it sluices around the sandbanks, in one estuary and out of another.

Stroomkarten exist for

1891 Westerschelde

1892 Oosterschelde
1893 Vokerak Grevelingen
1894 Haringvliet, Hollandsch Diep
1895 Hoek van Holland, Rotterdam
1896 Waddenzee

Books

OTHER NAUTICAL REFERENCE BOOKS

Reed's Nautical Almanack needs no introduction and is a mine of information useful to those who are going to make the North Sea passage to the Netherlands.

North Sea Harbours, Calais to Den Helder, by E. Delmar-Morgan, is the classic guide to all the sea harbours from Calais along the French and Belgian coasts and up the Dutch coast as far as Den Helder. It is an excellent guide to the sea-going cruising man.

Inland Waterways of the Netherlands, by E. E. Benest, is in three volumes and contains some general information about cruising Dutch inland waters and a vast amount of detail concerning widths, heights, depths, lengths, etc, of all the locks, bridges, rivers and canals of the Netherlands.

GENERAL CRUISING BOOKS ON HOLLAND

The following books make excellent general reading about boating and the waterways of Holland.

Holland Through Her Waterways, by Netherlands Yacht Builders and Yacht Owners Association.

Small Boat Through Holland, by Roger Pilkington.

'*Willynilley*' *to the Baltic*, by John Seymour.

GUIDE BOOKS

Of the guide books prepared with the general tourist in mind but which make excellent reading for the holiday cruiser, I recommend:

Fodor's Holland
Out and About in Holland by two Dutch retired schoolmasters, is a very concise Dutch publication written in English. As a ready reference book on what to see and where to go in any town you may cruise to, it is excellent.

GENERAL READING

The Netherlands, by Max Schuchart in the 'New Nations and People's Library' makes excellent background reading and I would advise anyone visiting Holland for whatever purpose and by whatever means to read this book.
The Dutch, How They Live and Work, by A. Hoffman is also very good background reading.

Those cruising skippers who are more at home in the open sea than on inland waters *must* read Charles Hadfield and Michael Streat, *Holiday Cruising on Inland Waterways*, to obtain that information which is relevant to inland waterways in all countries and which I have left out of this book.

How Safe is your Craft? is a Ministry publication and sets out the minimum safety requirements for small craft. It is obtainable from:

Information Division, Room 555,
Department of Trade and Industry,
1 Victoria Street, London SW1.

TOURIST INFORMATION

In every town in Holland that is of any size you will find an office of the VVV. From this organisation you can obtain street maps and general information about the town you are visiting. Many of the VVV offices change travellers' cheques and I have always found the staff employed most obliging and very helpful.

Information about Holland can be obtained in England from:

The Netherlands National Tourist Office,
143 New Bond Street,
London W1.

BRITISH CONSULATES

It is unlikely that you will require the services of the British Consulates, and in general holiday-makers are encouraged to sort out their own problems. There is, however, always the exceptional case and should you require their services they can be found at:

Herengracht 460, Amsterdam.
Lange Voorhout 10, 's-Gravenhage.
Parklaan 18, Rotterdam.

ANNUAL EVENTS IN HOLLAND

National holidays

1 January.
30 April, H.M. Queen Juliana's birthday.
25 and 26 December.

Other Events

May Scheveningen and Ijmuiden. Review of herring fleet.
Alkmaar. Cheese market (through to September).
Gouda. Cheese market (through to October).

June Middelburg. *Son et Lumière* (through to September).
Assen. TT Motor races (July also).
Arnhem. NATO Tattoo (July also).

July Warmond. Kaag week. Sailing races.
Loosdrecht. Sailing races (August also).
Zandvoort. Dutch Grand Prix.

August Sneek. Sailing races.
Delft. Tattoo.

FERRY SERVICES

British Rail. Day and night sailings from Harwich to Hoek van Holland. British Rail, Car Ferry Centre, Terminal House, 52 Grosvenor Gardens, London SW1 OAG.
North Sea Ferries Ltd. Departures daily (including Sundays) at 18.00 hours. Hull to Rotterdam. North Sea Ferries Ltd, King George Dock, Hedon Road, Hull, HU9 5QA.
Tor Line. Services between Grimsby and Amsterdam. Tor Line Ltd, West Gate, Immingham Dock, Nr. Grimsby, Lincs DN40 2PH.
Transport Ferry Services. Day and night crossings from Felixstowe to Rotterdam. Transport Ferry Service, 25 Whitehall, London SW1.

CHAPTER EIGHT

Glossary of Useful Words

Since very many Dutch people speak English you are unlikely to have to speak Dutch during your cruise. This glossary is therefore intended only to help you translate the signs you may see during your cruise. I have kept it short in order that you will not have far to look for the word you need.

Occasionally you may have to ask the way to a particular town, or in conversation you may wish to mention some place you have visited or wish to visit. For this reason I have included here a few pointers to help you pronounce place names in a way that is intelligible to the Dutch.

In general vowel and consonant sounds are as in English with the following exceptions.

aa as in ah!
ee as the *a* in shape.
ei as in height.
ie as the *ee* in sleep.
oo as the *oa* in road.
ou as in plough.
ij is like a *y* and is pronounced like the English *i* in right.
u is pronounced as in tube.
g is a strong guttural aspirate and can be pronounced like *h*.
r is slightly rolled.
v is like the English *f*.
d when terminal is pronounced like an English *t*.

Glossary of Useful Words

Dutch–English

aanleggen	to moor
achter	aft
achtersteven	stern
afgesloten	closed
afwisselenlicht	alternating light
alleen	only
als	as, if
anker	anchor
ankerlicht	riding light
baggermolen	dredger
bakboord	port (side)
baken	beacon
balkhelling	slipway
beneden	below
benzine	petrol
blauw	blue
bliksen	lightning
boeg	bows
boei	buoy
bolder	bollard
boord licht	navigation light
boot	boat
boven	above
brugeld	bridge dues
brulboei	whistle buoy
dekbalk	beam
diep	deep
diepgang	draught
Dinsdag	Tuesday
dok	dock
donder	thunder
Donderdag	Thursday
dood tij	neap tide
drijfbaken	spar buoy
droogdok	dry dock
duiker	diver
ebstroom	ebb tide
en	and
fluitboei	whistle buoy
geel	yellow

Glossary of Useful words

Dutch–English

geleidelicht	leading light
gevaar	danger
goed	good
groen	green
groepchitter	group flashing
groeponderbrokenlicht	group occulting light
groot	big
haven	harbour
havengeld	harbour dues
hoog	high
hoogwater	high tide
hijsen	hoist (verb)
huren	to rent
ingang	entrance
jacht	yacht
jol	dinghy
kaai	wharf
kaart	chart
keerdam	weir
klein	small
kogelton	spherical buoy
laag	low
laagwater	low water
lang	long
langszij ligger	lay alongside
langzam	slow
lengte	length
links	left
losgooien	cast off
maan	moon
Maandag	Monday
machine	engine
maximumsnelheid	speed limit
meeren	to moor
mooi weer	fine weather
naar	to
niet	not
nieuw	new

Dutch–English

oever	shoal, bank
olie	oil
onderbrokenlicht	occulting light
ondiep water	shallow water
oud	old
Pass op!	Attention!
petroleum	paraffin
pier	pier
rechts	right (direction)
reddingsboei	lifebuoy
reddingsboot	lifeboat
regen	rain
roer	rudder
rood	red
schitterlicht	flashing light
schutsluis	tide lock
slecht	bad
sleepboot	tug boat
sluis	lock
snel	speed
spitseton	conical buoy
spring-vloed	spring tide
stompeton	can buoy
stootkussen	fender
straat	strait, street
strijken	lower (verb)
stuurboord	starboard
te huur	to let
toplicht	masthead light
tweetakt	two-stroke
uit	out
uitgang	exit
vaargeul	channel
vanglijn	painter
varen	to sail
vastlicht	fixed light
vastmaken	make fast
veer	ferry
verboden	forbidden

Glossary of Useful Words

Dutch–English

verboden toegang	no admittance
verf	paint
viertakt	four-stroke
vlag	flag
vlot	raft
voetgangers	pedestrians
Vrijdag	Friday
vuurtoren	lighthouse
waarschuwing	warning
werk in uitvoering	repairs in progress
weilrijders	cyclists
windstil	calm
wit	white
Woensdag	Wednesday
Zaterdag	Saturday
zeearm	estuary
zeebrief	ship's papers
zon	sun
Zondag	Sunday
zwart	black
zwemvest	life-jacket

APPENDIX A

Weights and Measures

Conversions

Litres	*Gallons*
1	0·22
2	0·44
3	0·66
4	0·88
5	1·10
6	1·32
7	1·54
8	1·76
9	1·98
10	2·20

Gallons	*Litres*
1	4·5
2	9·1
3	13·6
4	18·2
5	22·7
6	27·3
7	31·8
8	36·3
9	40·9
10	45·4

Kilometres	*Statute Miles*
1	0·62
2	1·24
3	1·86
4	2·48
5	3·11
6	3·73
7	4·35
8	4·97
9	5·59
10	6·21

Statute Miles	*Kilometres*
1	1·6
2	3·2
3	4·8
4	6·4
5	8·1
6	9·7
7	11·3
8	12·9
9	13·5
10	16·1

1 nautical mile = 1·152 statute miles.

Metres	*Feet*
1	3·3
2	6·6
3	9·8
4	13·1
5	16·4
6	19·7
7	23·0
8	26·2
9	29·5
10	32·8

Feet	*Metres*
1	0·30
2	0·61
3	0·91
4	1·22
5	1·52
6	1·83
7	2·13
8	2:44
9	2.74
10	3·05

1 fathom = 6 feet.

Speed conversion graph

The speed limits on some of the inland waterways of the Netherlands are sometimes given in metres/minute and sometimes in kilometres/hour. Most British boats equipped with a log will have the log calibrated in either knots or mph. The graph is designed to enable the reader rapidly to interconvert any of these speed units, over the range of speeds most frequently used as speed limits.

To use the graph

(i) To convert metres/minute into either kilometres/hour, knots or mph.

Select the required speed in metres/minute on the vertical scale. Draw an imaginary line across the graph until it cuts the knots (mph or kilometres/hour) line. Project a line downwards from this point and read off on the horizontal scale the speed in knots (mph or kilometres/hour).

(ii) To interconvert knots, mph and kilometres/hour.

Example: To convert 6 knots to kilometres/hour.

Draw a line vertically from 6 on the horizontal scale to cut the knot line. From this point draw a horizontal line to cut the kilometre/hour line. From this point draw a vertical line to cut the horizontal scale. Read off the speed in kilometres/hour (11.2).

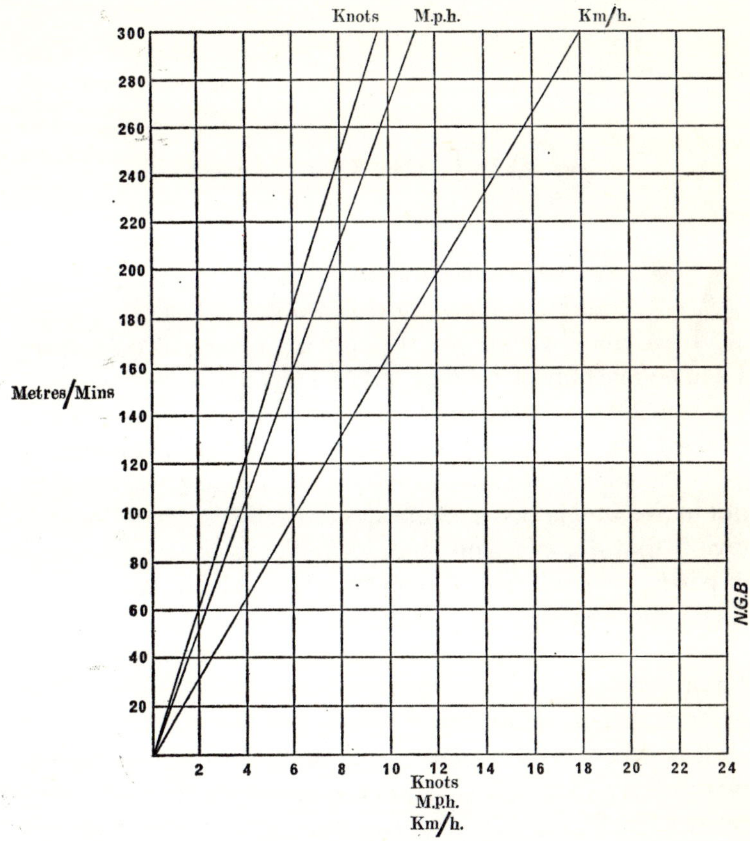

Speed conversion graph

APPENDIX B

Mast Lowering Gear

Many owners of sailing yachts find the difficulty of lowering the mast a real problem when cruising on inland waterways. During my first visit to Holland I had no mast lowering gear, but I had a look at many designs used on Dutch yachts and by a process of simplification I evolved the system set out below. Unlike some systems it gives the minimum clutter on the deck, is hardly visible when not in use, and is easy and cheap to install. Its only disadvantage is that it confers no lateral stability to the mast when it is partly lowered. The system can be modified to suit any single-masted yacht that has the mast stepped in a tabernacle.

Construction

A single length of ¾in water pipe is cut exactly in half, and each half bent to fit snugly inside the toerail. (If the pipe is placed through a hole in a thick piece of wood it will bend easily.) The forward ends of the pipe, which should come to a position on the deck vertically below the top of the bottle screw on the forestay, are fastened together by bolting a plate across them. This plate should have a U-bolt fixed in the middle.

The after ends of the pipe pass between pairs of right-angle plates about 3in long and about 2ins high, which are bolted (or screwed) to the deck. By fitting bolts in holes drilled through the ends of the pipes and the flanges of the plates, the pipes now in the shape of a wishbone, can be pivoted about their after ends.

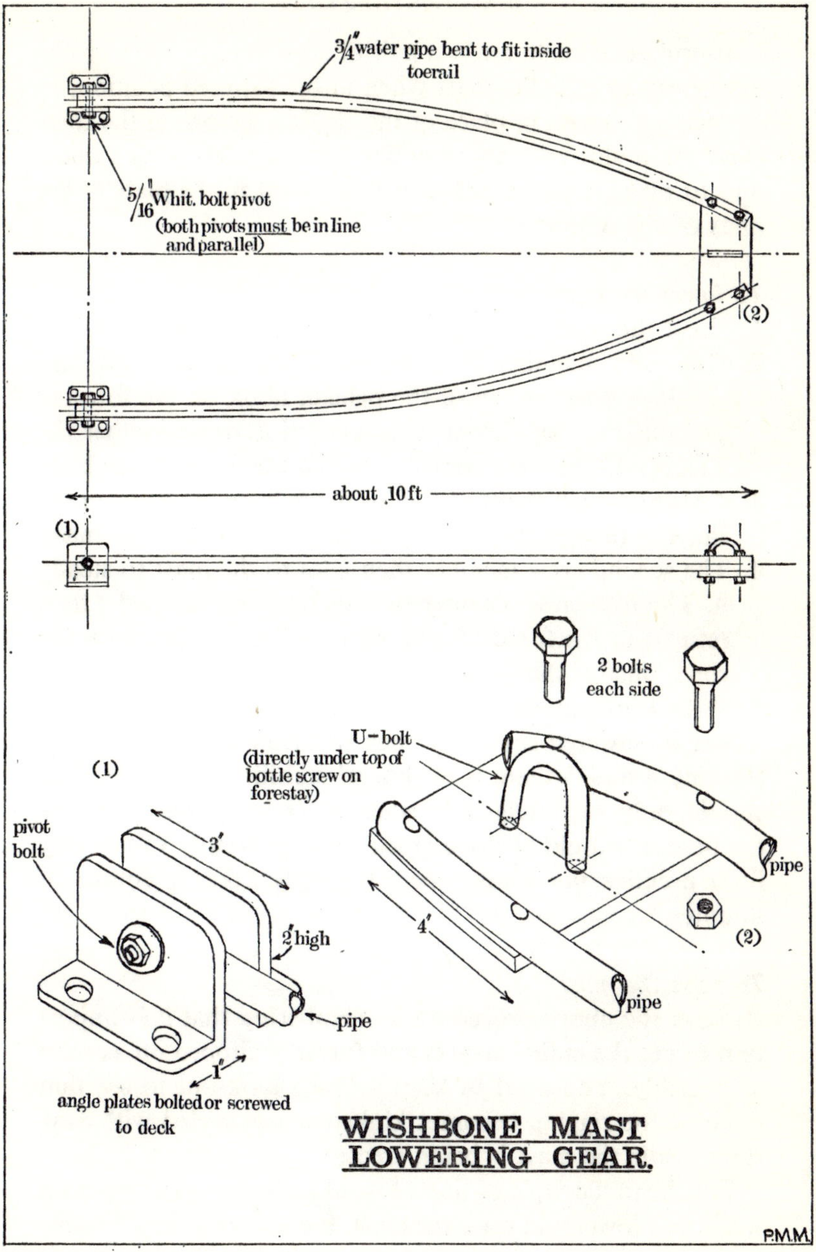

Mast lowering gear

It now remains to make a pair of sheer legs to erect on the after-deck to take the mast when in the lowered position, to fit 'pelican' hooks in place of the rigging screws on the forestay, main shrouds, and any shrouds forward of the mast, and to place a snap shackle in the U-bolt in the plate at the apex of the wishbone.

To lower the mast

1. Erect the sheer legs on the after-deck.
2. One crew member lifts the wishbone at the apex and shackles it to the top of the pelican hook on the forestay and passes a rope from the same point around either the stemhead roller or some other suitable fitting near the bow.
3. A second crew member unclips the pelican hooks from the shrouds (except the forestay) that are forward of the mast.
4. The second crew member then goes to the mast to steady it. The first crew member unclips the main shrouds, then returns to the foredeck and after taking the strain on the rope, unclips the forestay.
5. With a little guidance from the man at the mast, the latter can be lowered smoothly back to sit on the sheer legs.

(During 4 and 5 there is no lateral stability so this exercise should only be indulged in when there is no wash to rock the boat. It should be done quickly and smoothly.) With very little practice the whole operation only takes a couple of minutes.

To raise the mast

Reverse the above procedure, remembering that it is important to get the main shrouds and forestay clipped on as soon as possible. You need to keep a sharp look-out to see that none of the rigging becomes kinked or entangled with anything while the mast is being erected.

For the first couple of times I would advise you to moor up while you lower and raise the mast, but you will soon be able to do it under way.

Acknowledgements

It would not have been possible for me to write *Holiday Cruising in the Netherlands* had it not been for the help and co-operation of many people both in England and the Netherlands. In particular I am grateful for the help given me by the Netherlands National Tourist Office in London and the ANWB in Amsterdam. My sincere thanks go to Captain Armand E. Westzaan of the Netherlands Merchant Navy and to Mr Colin S. Craig for translating the Dutch canal regulations into English for me, and to Captain W. T. Rodger, a Trinity House pilot, for his help and advice on the buoyage system of the Netherlands.

I am very grateful to four of my sixth-form students. To Paul Milton for his drawing of the mast lowering gear; to Mark Evans, Ian Brown and Nick Brown for their work on the sketch maps; and to the eleven students of mine who at one time or another have crewed for me on my trips to the Netherlands.

My thanks go to Power Marine, Kes Weber, and the Jachthaven Neptunus for the loan of a power boat in which to explore parts of the great rivers. To the harbourmaster of the KNZ en RV (Royal Netherlands Yacht Club) for looking after *Puffdragon* while I was away. Lastly my thanks go to the many British and Dutch yachtsmen who have made valuable suggestions and comments that have helped me in writing this book.

Index

When several page numbers are shown against one entry, principal references in bold type should be consulted first.

Afluitdijk, 45
Akersloot, 69
Alkmaar, 79
Alkmaardermeer, **69**, 79
Alphen a/d Rijn, 73
Ameland, 66
Amstelmeer, 69–70
Amstel River, 73
Amsterdam, 73, **78**; from Rotterdam, 95; to Den Helder, 96; to Friesland, 96
Amsterdam–Rijnkanaal, 76; crossing of, 87–8; *caution*, 98
Angling, 103–4
Antwerp, 28
Arnhem, 86
Assen, 60
Avifauna, 73

Balk, 59
Barges, 102
Belfeld, 92
Belterwijde, 52
Bergen-op-Zoom, 29
Bergumermeer, 55
Beulakerwijde, 52
Biesbosch, 92
Bolbaken, 85
Bolsward, 59
Boskoop, 73
Braassemermeer, 70
Breda, 94
Breskens, 28
Brewster, William, 75
Bridges, 12; signals, **112**
Brielle (Brill), 39
Brouwershaven, Zeegat van, 31
Bruinisse, 31

Castles, International Institute of, 86
Catamarans, 24, 42
Cats, Jacob, 31
Charlemagne, 91
Colijnsplaat, 30
Customs: Netherlands, 22; UK, 12

De Cocksdorp, 63
De Hoorne, 69
De Houkes, 70
De Koog, 63
De Leyen, 55
Delft, 74
Delftware, 46, **74**
Delfzijl, 47, **61**
Den Burg, 63
Den Helder, 79–80
Deventer, 85
Dinghy cruising, 16; in Friesland, 49
Doesburg, 86
Dokkum, 61
Donge, 94
Dordrecht, 28, **34–5**
Dordtsche Kil, 34

Drenthe, 52
Durgerdam, 42

Edam, 43
Ee, 59
Eemdijk, 48
Eemskanaal, 56
Eendracht, 29
Eerste Friese Zeilschool, 57
Eindhoven, 93
Eindhovenskanaal, 93
Eirland, 63
Elburg, 48
Electrical equipment, 15
Emmeloord, 60
Enkhuizen, 43
Europort, 37

Fluessen, 55
Follega Sloot, 57
Friesland, 19, **49–62**; canals of, 58–62
Frisians, 49–50

Gas, 15
Geijsteren, 92
Ghentkanaal, 29
Gierpont, **84**, 87
Goes, 29
Gorinchem, 91
Gouda, 73
Grave, 92
's-Gravenhage, 74–5
Greate Pier, 61
Grevelingen, 28
Grevelingendam, 31
Groeve, 56
Groningen, 52, **61**
Grote Brekken, 57, 60
Grote Gaastmeer, 55
Grote Zijlroede, 59
Grouw, 56

Haarlem, 75
Hals, Frans, 75
Hansweert, 29
Harderwijk, 48
Haringvliet, 28, **33**
Harlingen, 60
Hattem, 85
Heeg, **55**, 59
Heegermeer, **55**, 59
Hellevoetsluis, 33
's-Hertogenbosch, 91
Hindeloopen, 46
Hoek van Holland, 28
Hollandsch Diep, 28, **34**
Hoorn, 43
Houkesloot, 57
Huizen, 48

Ijlst, 59
Ijmuiden, 77
Ijssel, 48, 81, **84–6**
Ijsselmeer, 39–42; from Rotterdam, 76

Julianakanaal, 93

Kabelpont, 84
Kagerplassen, 70
Kampen, 84–5
Ketelmeer, 60
Ketelmond, 48
Kibbaken, 85
Krabbenkreek, 31
Krammer, 31
Krimpen a/d Ijssel, 73
KRZV de Maas (Royal Maas YC), 38
Langweer, 59
Lauwersmeer, 61
Leeuwarden, 60
Leiden, 70, **75**
Lek, River, 76
Lekkanaal, *caution*, 98
Lely, Dr C., 45
Lelystad, 47
Lemmer, **46–7**, 57, 60; to Delfzijl, 97
Limburgse Maas, 92–3

Locks, 12; signals, 112; staircase, 94; at Willemstad, 31–3

Maas, River, 38, 81, **91**
Maasbracht, 93
Maassluis, 38
Maastricht, 93
Makkum, **45–6**, 59
Marken, 42–3
Medemblik, 44
Meppel, 52, 60
Meppeler, Diep, 52
Merwedekanaal, 76
Middelburg, 26–7
Middelharnis, 33
Moerdijk, 34
Monnickendam, 42
Morra, 55
Muiden, 47–8
Museums: Dutch Railway, 77; Friesian, 60; Netherlands Open Air, 86; West Friesian, 43; Zuyderzee, 44

Nature reserves: Veersemeer, 27; Waddenzee, 63
Nauwe Wijmerts, 59
Neder Rijn, 86–7
Nieuwe Maas, 28, 37
Nieuwe Waterweg, 37
Nijkerk, 48
Nijmegen, 91
Noord, 37
Noorder Oudweg, 59
Noordholland, 66ff
Noordhollandskanaal, 69, **78–80**
Noordoost Polder, 47, 60
Noordwest Overijssel, 52
Noordzeekanaal, 77–8
North Sea, crossing, 14

Oosterschelde, 28, **29–30**
Oss, 91
Ouddorp, 31
Oude Karre, 59
Oude Maas, 28, **37**
Oudergaster Brekken, 55
Oudeschild, 62
Overijssel, 52

Pampushaven, 47
Pannerdenskanaal, 88
Philips, Gerard, 93
Pikmeer, 56
Pilgrim Press, 75
Prinsenhof, 56
Prinses Margrietkanaal, 52, 60
Purmer, 79
Purmerend, 80

Radboud Castle, 44
Randmeren, 48
Rembrandt, 60, 75
Roderbrug, 92
Roermond, 92
Rotterdam, 37–8; from Dordrecht, 34, **37–8**; from Zeeland, 95; to Amsterdam, 73, 95–6; to Ijmuiden, 76
Rotterdamsche Waterweg, *see* Nieuwe Waterweg
Rozendaal, 86
Rudolph Agricola, 61

Safety, 16
St Annaland, 31
St Anna Parochie, 60
Saskia van Vylerburgh, 60
Sassenheim, 70
Scharster Nieuwe Rijn, 59
Schiermonnikoog, 66
Schildmeer, 56
Schokland, 47
Schoonhoven, 76
Schouwen Duiveland, 30
Slipping, 16
Sloten, 57
Slotermeer, **57**, 59
Sneek, **57**, 59
Sneekermeer, **57**, 59
Spui, 33
Sijtebuurster Ee, 56

Staveren, 46
Stolpen, 80
Strijensas, 34
Swimming, 103–4

Terneuzen, 28–9
Terschelling, 65
Texel, 62–3
Tholen, 29, 30
Tiel, 91
Tilburg, 94
Tjeukemeer, **57**, 59
Trawlers, 103

Uitgeest, 69
Uithoorn, 73
Urk, **47**, 60
Urkervaart, 60
Utrecht: city, 76–7; province, 66ff

Vafamil, Jachthaven, 70
Van Harinxmakanaal, 60
Van Panhuijsbrug, 59
Van Panhuijskanaal, 59
Vecht, River, 76
Veere, 27
Veersemeer, **27–8**, 29
Vlakke Brekken, 55
Vlieland, 63
Vlissingen, **24–5**, 28
Volendam, 42
Volkerak, 28, 31
Vreeswijk, 76
Vrouwentroost, 70

Waal, River, 81, **88–91**
Waddenzee, 62
Waddinxveen, 73
Walcheren, Kanaal door, 24
Wanssum, 92
Water Skiing, 16
Westeinderplassen, 70
Westerschelde, 28–9
Wilhelminakanaal, 94
Willemsdorp, 34
Willemstad, 34; locks, 31–3
Workum, 59
Wormer, 79
Woudsend, 57, 59
Wijde of Peanster Ee, 56

Zaandam, 77
Zandkreek, 29
Zandkreekdam, 29
Zandmeer, 55
Zeelandbrug, 29, 30
Zierikzee, 30; *caution*, 98
Zuid Beveland, Kanaal door, 29
Zuidholland, 66ff
Zuidlaardermeer, 58
Zuidwillemsvaart, 93
Zutphen, 86
Zuyder Zee, *see* Ijsselmeer
Zwartemeer, 52
Zwartsluis, 60
Zwolle, 85
Zijpe, 31